PRACTICAL PROBLEMS in MATHEMATICS

for HEATING AND COOLING TECHNICIANS

Second Edition

PRACTICAL PROBLEMS in MATHEMATICS

for HEATING AND COOLING TECHNICIANS

Second Edition

RUSSELL DEVORE

DELMAR PUBLISHERS inc. ®

NOTICE TO THE READER

Publisher does not warrant or guarantee any of the products described herein or perform any independent analysis in connection with any of the product information contained herein. Publisher does not assume, and expressly disclaims, any obligation to obtain and include information other than that provided to it by the manufacturer.

The reader is expressly warned to consider and adopt all safety precautions that might be indicated by the activities described herein and to avoid all potential hazards. By following the instructions contained herein, the reader willingly assumes all risks in connection with such instructions.

The publisher makes no representations or warranties of any kind, including but not limited to, the warranties of fitness for particular purpose or merchantability, nor are any such representations implied with respect to the material set forth herein, and the publisher takes no responsibility with respect to such material. The publisher shall not be liable for any special, consequential or exemplary damages resulting, in whole or in part, from the readers' use of, or reliance upon, this material.

Delmar staff

Executive Editor: David Gordon
Project Editor: Judith Boyd Nelson
Production Coordinator: Wendy Troeger
Design Supervisor: Susan Mathews

For information, address Delmar Publishers Inc.
3 Columbia Circle, PO Box 15015
Albany, New York 12212-5015

Printed in the United States of America
Published simultaneously in Canada
by Nelson Canada,
A division of The Thomson Corporation

10 9 8 7 6 5 4

CONTENTS

PREFACE

The student learning heating and cooling theory must be familiar with the terminology and practices of heating and cooling technicians. PRACTICAL PROBLEMS IN MATHEMATICS FOR HEATING AND COOLING TECHNICIANS, second edition, provides the student with the practical and realistic mathematical problems that are encountered by heating and cooling technicians. By solving the problems, the technical and mathematical aspects are both strengthened, thus providing a solid foundation for a career as a heating and cooling technician.

PRACTICAL PROBLEMS IN MATHEMATICS FOR HEATING AND COOLING TECHNICIANS, second edition, is one in a series of workbooks that can be used in conjunction with a comprehensive mathematics textbook—such as BASIC MATHEMATICS SIMPLIFIED by C. Thomas Olivo and Thomas P. Olivo (Delmar Publishers Inc.)—or in an individualized math program for students with interests in specific fields. Each workbook provides practical experiences in using mathematical principles to solve occupationally related problems.

This series of workbooks is designed for use by a wide range of students. The workbooks are suitable for any student from the junior high school level through high school and up to the two-year college level. These workbooks, though designed for the vocational student, are equally suitable for the liberal arts student.

The series has many benefits for the instructor and for the student. For the student, the workbooks offer a step-by-step approach to the mastery of essential skills in mathematics. Each workbook includes relevant and easily understandable problems in a specific vocational field.

For the instructor, the series offers a coherent and concise approach to the teaching of mathematical skills. Each workbook is complemented by an instructor's guide that includes answers to every problem in the workbook and solutions to many of the problems. The instructor's guide also includes the diagnostic reading survey as well as references to comprehensive mathematics textbooks published by Delmar Publishers. In addition, three achievement reviews are provided at the end of each workbook to provide an effective means of measuring the student's progress.

Both the instructor and the student will benefit from the specific vocational material and the appendix materials such as performing the basic operations with denominate numbers.

The author was an instructor in the mathematics and physics division at Bloomsburg State College in Pennsylvania and is presently teaching at a training center for the Pennsylvania Power and Light Company. Dr. DeVore was named in the 1976 editions of Who's Who in South Carolina and Outstanding Young Men of America.

Other workbooks in this series are:

PRACTICAL PROBLEMS IN MATHEMATICS FOR

AUTOMOTIVE TECHNICIANS
CARPENTERS
CONSUMERS
COSMETOLOGY
ELECTRICIANS
GRAPHIC ARTS
HEATING AND COOLING TECHNICIANS
MACHINISTS
MASONS
MECHANICAL DRAFTING
PRINTERS
SHEET METAL TECHNICIANS
THE METRIC SYSTEM
WELDERS

USING A CALCULATOR

Today the availability and low cost of calculators have made many math problems less of a chore. The modern calculators can do many calculations quickly. Thankfully, a complicated math calculation is no longer a big problem.

However, calculators can also have a "bad" side to them. People rely on them to do all of their math; they have gotten lazy when doing math. These people may get into trouble for a couple of reasons. First, when they are doing a math problem, these people have no idea if they have done it correctly. If an incorrect number was entered, or an incorrect button was hit, these people would never know because they have no idea what kind of answer to expect. A couple of examples might clarify. A person who relys totally on the calculator for math problems is multiplying 8.4 × 2.1. The calculator gives 4 as the answer. This person does not realize that he made a mistake (pressing the ÷ button instead of the × button), because he does not know that the correct answer should be about 16. A person doing a problem may not realize that a decimal point was not pressed or that one zero did not enter the calculator correctly when the number was being entered, because, once again, he has no idea what the correct answer should be. So multiplying 8.4 × 2 may give an answer of 168 rather than 16.8 and the person doing the problem may not know that an error had been made.

There is a second drawback to calculators: They cannot work in fractions. If a person relies on the calculator all of the time, he will have no idea how to solve the problem because he has not practiced. Solving math problems is largely a matter of practicing; without practice, problems are hard to solve.

Yet even with these drawbacks, calculators can be a big help in math. They can solve all problems with decimals and are really helpful when there are many numbers to add, subtract, or multiply. It is important to follow some basic rules.

1. Get in the habit of thinking about what a reasonable answer would be. This makes it easier to notice when an error has been made.

2. After entering the number into the calculator, check to see that you entered the correct number before pushing the operation button.

If you are going to buy a calculator, you will probably not need many of the special keys that some calculators have on them. They would just be a waste of money. More keys do not mean that the calculator is better or more accurate.

Each calculator comes with an instruction booklet. Use it. Read it completely. This is the only way that you will know exactly what your calculator will or will not do. One of the biggest shortcuts when working with most calculators is that when doing multiple operations, such as adding many numbers, you do not have to press the = button after each number is entered. Pressing the + sign will cause the calculator to add the number just entered to the total already in the calculator and be ready to have another number entered. You only have to press the = button once, at the very end to get the final sum. This same idea works for subtracting and multiplying. Remember that not all calculators work this way. Check with the instruction booklet to be sure.

As far as this book is concerned, it is suggested that *all* of the problems be worked by hand to get as much mathematical practice as possible. Some of the problems can then be checked by using the calculator to give practice in using the calculator. Throughout each unit, there will be a special indication in front of some math problem numbers. These are problems where a calculator could be used to check the problem.

Whole Numbers

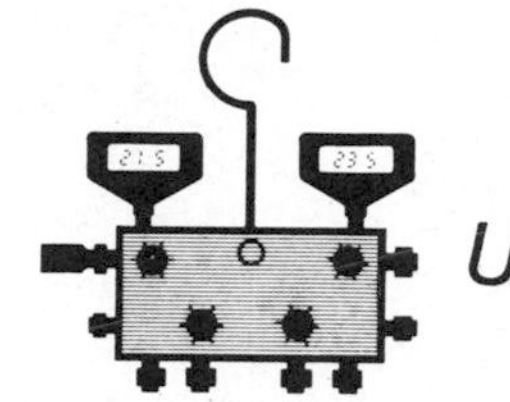

Unit 1 ADDITION OF WHOLE NUMBERS

BASIC PRINCIPLES

- Review and apply the principles of addition of whole numbers to the problems in this unit.
- Study addition of denominate numbers in section I of the appendix.

When a number is written, each numeral holds a certain place and those places have names. For example, the number 4 567 890 has 0 in the units place, 9 in the tens place, 8 in the hundreds place, 7 in the thousands place, 6 in the ten thousands place, 5 in the hundred thousands place, and 4 in the millions place. Whole numbers are numbers that have nothing smaller than a unit; they have no fractional part to them.

Addition is the process of finding the total (sum) of two or more numbers.

When adding whole numbers, it is best to place the numbers in columnar form. To do this, line the units places of each number underneath each other as shown below:

$$\begin{array}{r} 8 \\ 27 \\ 144 \\ 6 \\ \underline{97} \end{array}$$

By lining up the units, all of the other places are also lined up.

Now add the units column.

$$\begin{array}{r} 8 \\ 27 \\ 144 \\ 6 \\ \underline{97} \\ 2 \end{array}$$

If the total is greater than ten, carry the second number to the top of the tens column and then add that along with the tens column.

$$\begin{array}{r} \mathbf{3} \\ 8 \\ 27 \\ 144 \\ 6 \\ \underline{97} \\ 82 \end{array}$$

Do the same for each column to get the final answer.

$$\begin{array}{r} \mathbf{13} \\ 8 \\ 27 \\ 144 \\ 6 \\ \underline{97} \\ 282 \end{array}$$

PRACTICAL PROBLEMS

Add.

1. 625
+ 351

2. 2 273
5 402
+ 313

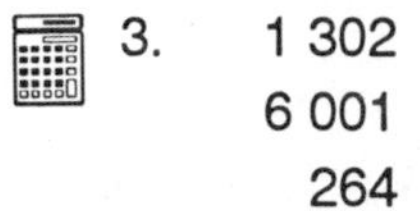

3. 1 302
6 001
264
+ 1 412

4. 129 square inches
453 square inches
+ 287 square inches

5. 567 meters
180 meters
35 meters
+ 208 meters

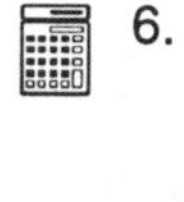

6. 3 413 gallons
9 020 gallons
787 gallons
4 121 gallons
+ 2 568 gallons

7. 247 + 501 + 131 __________

8. 632 + 51 + 3 __________

9. 7 204 + 60 + 13 + 21 __________

10. 656 grams + 804 grams + 222 grams __________

11. 1 717 feet + 485 feet + 9 204 feet + 339 feet

12. 13 minutes + 491 minutes + 877 minutes __________

13. This house is heated by electric baseboard heaters. Find, in feet, the total length of heaters needed. __________

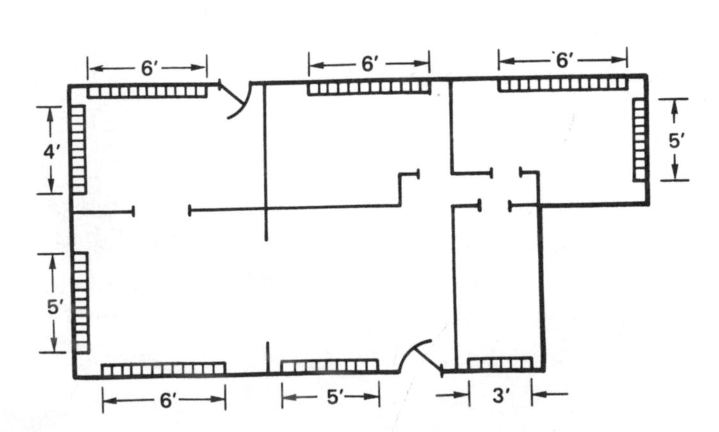

14. A forced air system takes return air at 62°F. It heats the air 59°F. What is the temperature of the heated air blown out to the room?

15. Find, in feet, the total distance air travels through this duct.

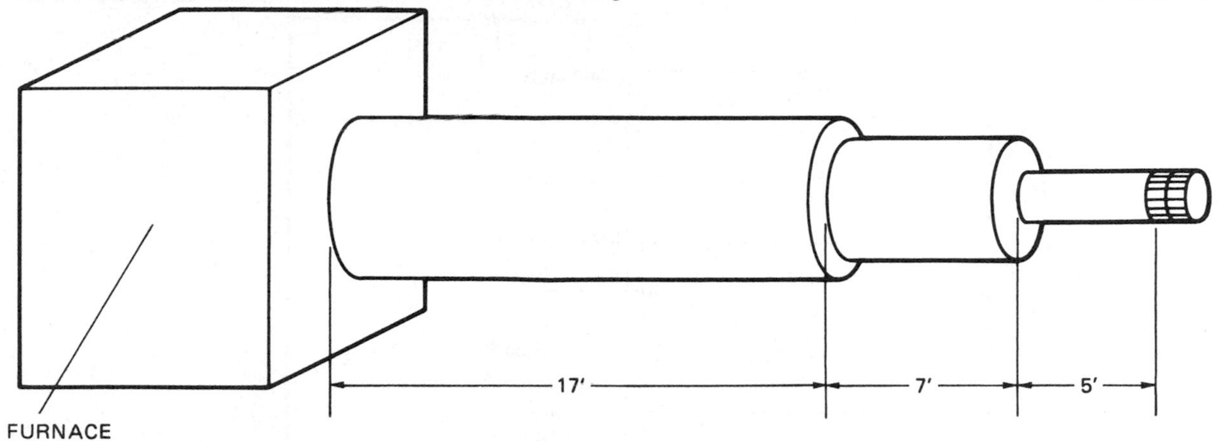

16. Pipe lengths of 2 inches, 8 inches, 108 inches, 8 inches, and 24 inches are used to make a conduit for wires to a heat pump. How many inches of pipe are needed for this conduit?

17. The condensing unit of an air conditioner is wired to the power source. The wires run along the wall of the house. Find, in feet, the amount of wire that is used.

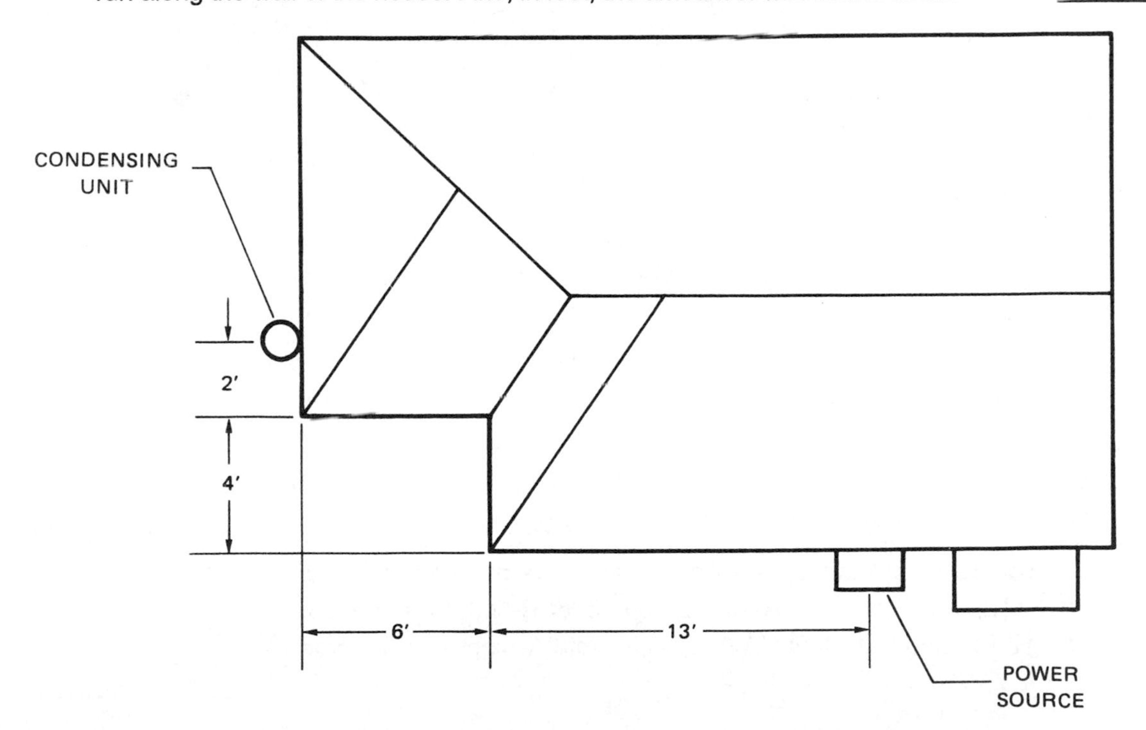

18. In one month, a repairer works 43 hours, 41 hours, 40 hours, and 41 hours. What is the total number of hours the repairer works?

19. Find, in cubic feet, the total volume of this house.

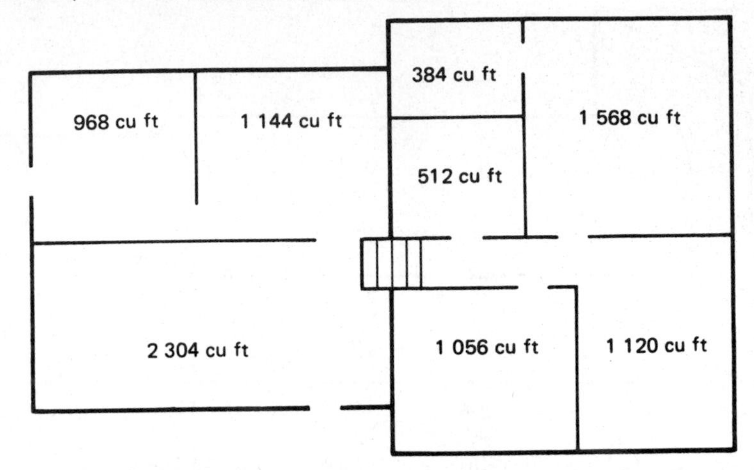

20. Starting from the shop, a repairer makes 5 stops in one day. Find, in miles, the total distance the repairer travels.

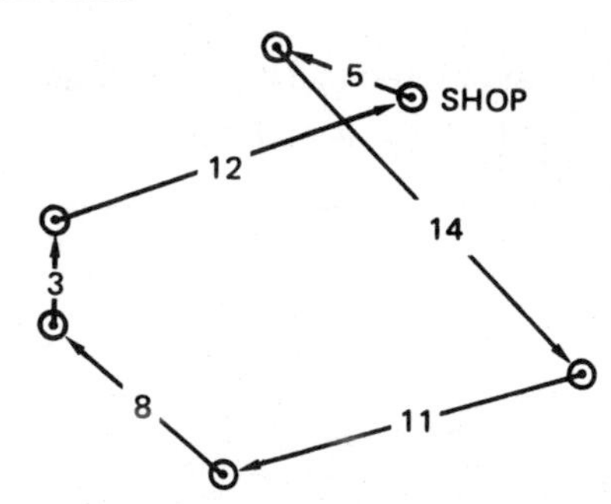

21. A forced convection heating system is installed in a house. The lengths of 6-inch circular duct that are needed are: kitchen, 6 feet; dining room, 12 feet; living room, 3 feet; master bedroom, 5 feet; second bedroom, 7 feet; third bedroom, 13 feet. How many feet of 6-inch round duct are needed?

22. A repair truck carries 121 feet of 3/4-inch copper tubing, 43 feet of 1/8-inch copper tubing, 76 feet of 1-inch conduit pipe, and 112 feet of 1/2-inch PVC pipe. What is the total length of material on the truck?

23. An air conditioning shop orders the following amounts of refrigerant: 125 pounds of R-112, 150 pounds of R-14, 70 pounds of R-502, 90 pounds of R-717, and 120 pounds of R-744. What is the total weight of refrigerant ordered?

24. Truck 1 carries 7 1-inch stop valves. Truck 2 carries 5 1-inch stop valves and truck 3 carries 8 1-inch stop valves. The warehouse has 19 1-inch stop valves. What is the total inventory for 1-inch stop valves? __________

25. During 1988, the Keep Kool Company's 5 repair trucks covered 7 252 miles, 8 917 miles, 4 266 miles, 7 793 miles, and 8 214 miles. What is the total mileage the Keep Kool Company should report for 1988? __________

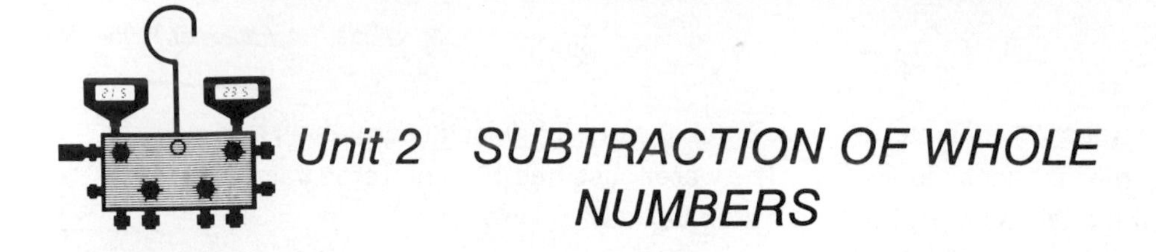

Unit 2 SUBTRACTION OF WHOLE NUMBERS

BASIC PRINCIPLES

- Review and apply the principles of subtraction of whole numbers to the problems in this unit.
- Study subtraction of denominate numbers in section I of the appendix.

Subtraction is the process of finding the difference between two numbers.

When subtracting, place the smaller number underneath the larger one and line up the units columns. The other columns will also be aligned. Beginning with the units column, subtract the lower number from the upper one.

```
 1749
– 563
-----
    6
```

It is not possible to subtract 6 from 4, so "borrow" 1 from the 7, making the 7 a 6. The 1 becomes a 10 and is added to the 4 making the 4 a 14. Now 6 can be subtracted from 14.

```
the 7 becomes a 6
     4 becomes 14
 1749
– 563
-----
 1186
```

PRACTICAL PROBLEMS

Subtract.

1. 87 – 23

2. 557 – 125

3. 927 – 216

4. 481 kilometers – 328 kilometers

5. 1 735 quarts – 466 quarts

6. 904 yards – 149 yards

7. 853 – 522 __________

8. 489 – 169 __________

9. 796 – 81 __________

10. 526 cubic feet – 440 cubic feet __________

11. 748 square centimeters – 159 square centimeters __________

12. 1 104 seconds – 378 seconds __________

13. This indoor-outdoor thermometer shows the temperature readings for a winter day. What is the temperature difference? __________

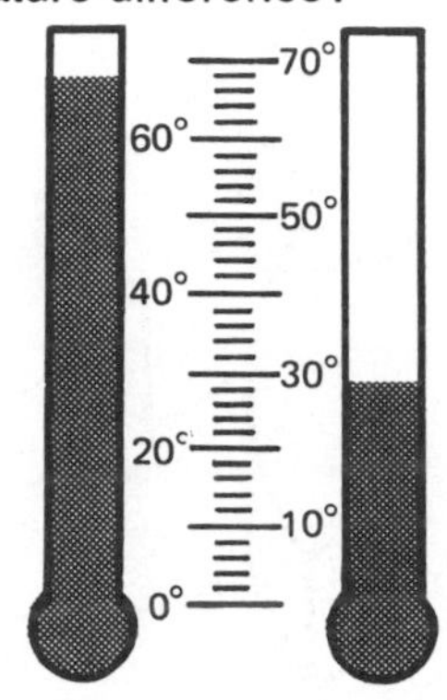

14. It takes 104 hours to install a heating system in an office building. The installers have already worked 36 hours. How many more hours are needed to finish the installation? __________

15. A cylinder containing refrigerant R-11 weighs 22 pounds. When empty, the tank weighs 4 pounds. How many pounds of R-11 are in the cylinder? __________

16. This room needs 19 feet of electric baseboard heaters. Lengths of 6 feet and 7 feet are already placed. What length of baseboard heater still must be placed? __________

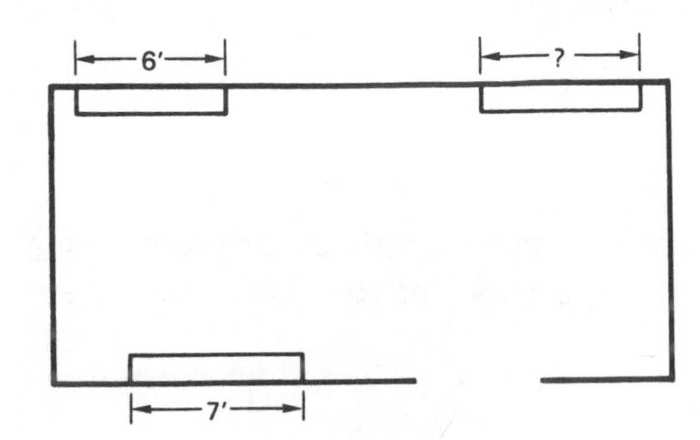

17. Air flows through the main duct of this system at a rate of 220 cubic feet per minute. In one of the branches, the rate is 96 cubic feet per minute. Find, in cubic feet per minute, the rate that the air flows through the other branch. ____________

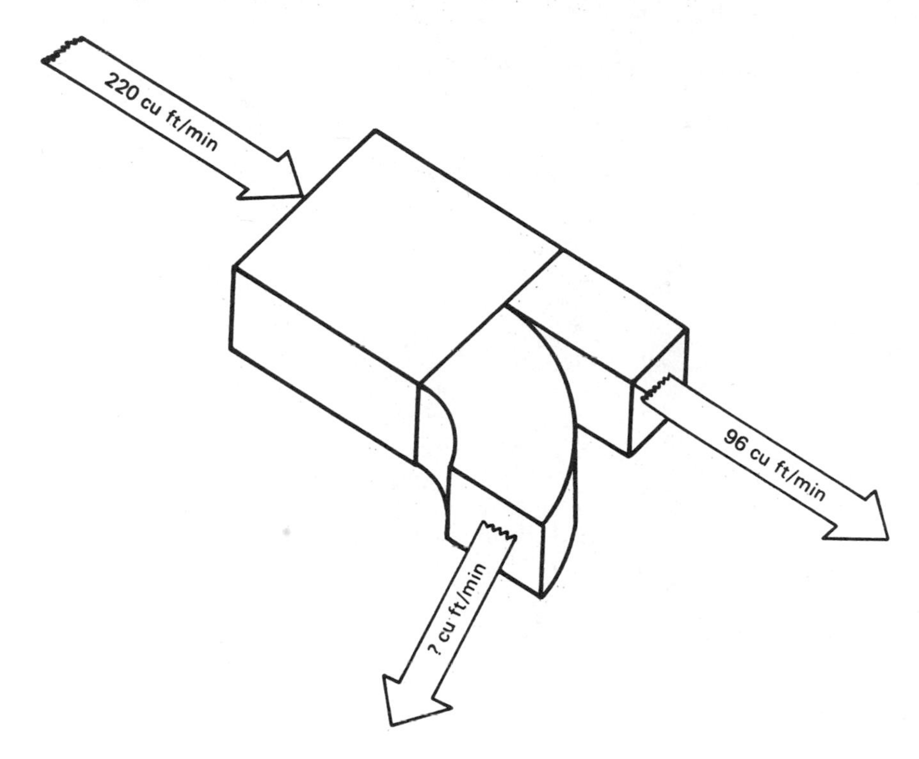

18. A heating and air conditioning repair shop keeps 1 500 feet of 3/4-inch copper tubing in stock. An inventory shows that there are only 623 feet on hand. How many feet of tubing need to be ordered? ____________

19. Three electric heaters draw 16 amperes of current. The first heater draws 6 amperes and the second draws 4 amperes. How many amperes are drawn by the third heater? ____________

20. A 162-foot roll of *#10* wire is used when installing a residential air conditioning unit. Lengths of 17 feet and 18 feet are cut from the roll. How many feet are left? ____________

21. This condensing coil has 3 grill openings. The total length of the openings is 57 inches. Find the length of the center opening. ______

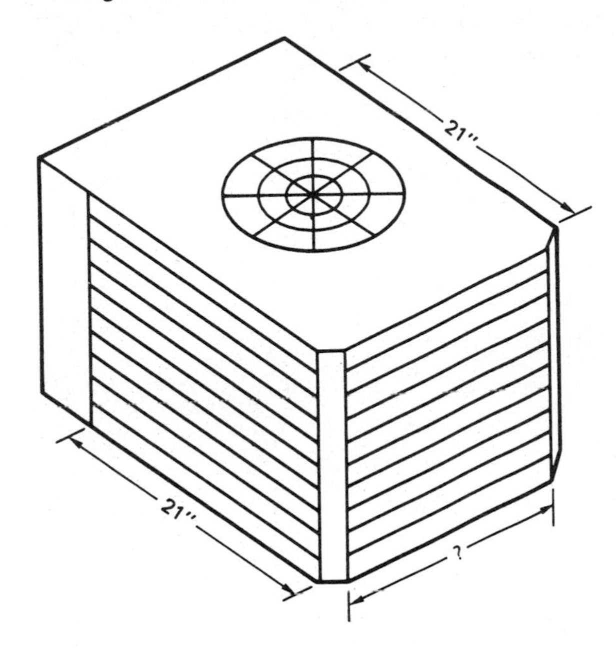

22. A full oil tank holds 280 gallons of Grade 2 fuel oil. During 4 months, the amounts of fuel oil used are: 0 gallons, 18 gallons, 53 gallons, and 123 gallons. How much fuel oil is left in the tank? ______

23. An installer has 84 strap hangers in the truck. On a job, 32 straps are used on hot water supply lines and 29 on return lines. The installer also uses 15 straps on an electrical conduit which carries wires to the furnace. Find the number of straps that are left. ______

24. An air conditioning installer orders 150 sheets of 3-foot by 8-foot sheet metal. The installer keeps this record.

NUMBER OF SHEETS	USE
13	6-inch circular ducts
19	4-inch circular ducts
25	10-inch x 12-inch rectangular ducts
8	3-inch x 5-inch rectangular ducts

How many sheets of metal are left? ______

25. A main air supply plenum for a room supplies air to 4 diffusers. The plenum has a flow of 150 cubic feet per minute (cu ft/min). Diffuser 1 has a flow of 45 cu ft/min, diffuser 2 has a flow of 40 cu ft/min, and diffuser 3 has a flow of 25 cu ft/min. What is the air flow through the fourth diffuser? ____________

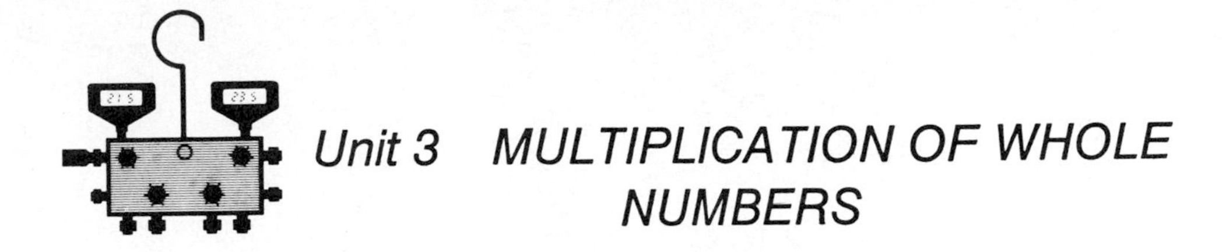

Unit 3 MULTIPLICATION OF WHOLE NUMBERS

BASIC PRINCIPLES

- Review and apply the principles of multiplication of whole numbers to the problems in this unit.
- Study multiplication of denominate numbers in section I of the appendix.

When multiplying two numbers, place one under the other and line up their units columns.

$$\begin{array}{r} 651 \\ \underline{\times 47} \end{array}$$

Start with the unit number of the second number and multiply it by the entire first number. Carry the number for any products larger than ten.

$$\begin{array}{r} \mathbf{3} \\ 651 \\ \underline{\times 47} \\ 57 \end{array} \qquad \text{(multiply 6 times 7 and then add 3)} \qquad \begin{array}{r} \mathbf{3} \\ 651 \\ \underline{\times 47} \\ 4557 \end{array}$$

Then multiply the tens number of the second number by the entire first number. Line the first number that you write down directly under the same column that you are multiplying by; in this case the number would be directly under the tens column.

$$\begin{array}{r} \mathbf{2} \\ 651 \\ \underline{\times 47} \\ 4557 \\ 2604 \end{array}$$

Finally add the numbers that you have written down, column by column. This will give you the final answer. (Notice that the answer is written with spaces at the proper places. This should be done while the answer is being written.)

$$\begin{array}{r} 651 \\ \underline{\times 47} \\ 4557 \\ \underline{2604} \\ 30\ 597 \end{array}$$

PRACTICAL PROBLEMS

Multiply.

1. 73 × 3

2. 41 × 29

3. 662 × 706

4. 56 inches × 9

5. 359 millimeters × 48

6. 499 gallons × 376

7. 19 × 65

8. 478 × 41

9. 509 × 92

10. 840 liters × 760

11. 321 square feet × 251 × 6

12. 420 × 537 × 86 hours

13. There are 144 electrical connectors in a box. How many connectors are there in 7 boxes?

14. An apartment building has 8 apartments in it. Each apartment has 7 duct openings which need diffusers. How many diffusers are needed for the entire building?

15. A repairer charges $9 per hour for labor. How much should be charged for a 17-hour job?

16. A can of Refrigerant 12 (R-12) for car air conditioners contains 14 ounces. A case of 24 cans has a total of how many ounces of R-12?

17. There are 36 cylinders of refrigerant R-11 in a stockroom. Each cylinder contains 137 pounds of refrigerant. How many pounds of R-11 are in the stockroom?

18. Using one 3-foot by 8-foot piece of sheet metal, an installer makes four 6-inch circular ducts. Each duct is 3 feet long. How many ducts can be made from 26 sheets of metal?

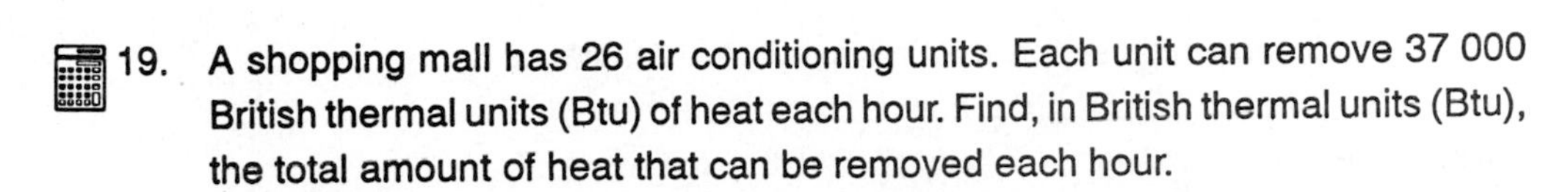

19. A shopping mall has 26 air conditioning units. Each unit can remove 37 000 British thermal units (Btu) of heat each hour. Find, in British thermal units (Btu), the total amount of heat that can be removed each hour. ________

20. Burning 1 gallon of Grade 1 fuel oil produces 137 000 Btu of heat. How many British thermal units of heat are produced when 250 gallons are burned? ________

21. One roll of duct insulation can insulate 25 linear feet of a duct. How many linear feet of a duct can be insulated with 73 rolls? ________

22. *Latent heat of fusion* is the heat given up as a substance freezes. The latent heat of fusion for water is 144 British thermal units per pound (Btu/lb). This means that 144 Btu are given up for each pound of water that freezes. How many British thermal units of heat are given up when 4 000 pounds of water freezes? ________

23. In an air conditioning repair shop, more storage space is needed. To make space, seven units of shelves are purchased. Each 5-shelf unit is 6 feet long. How many feet of shelf space are added? ________

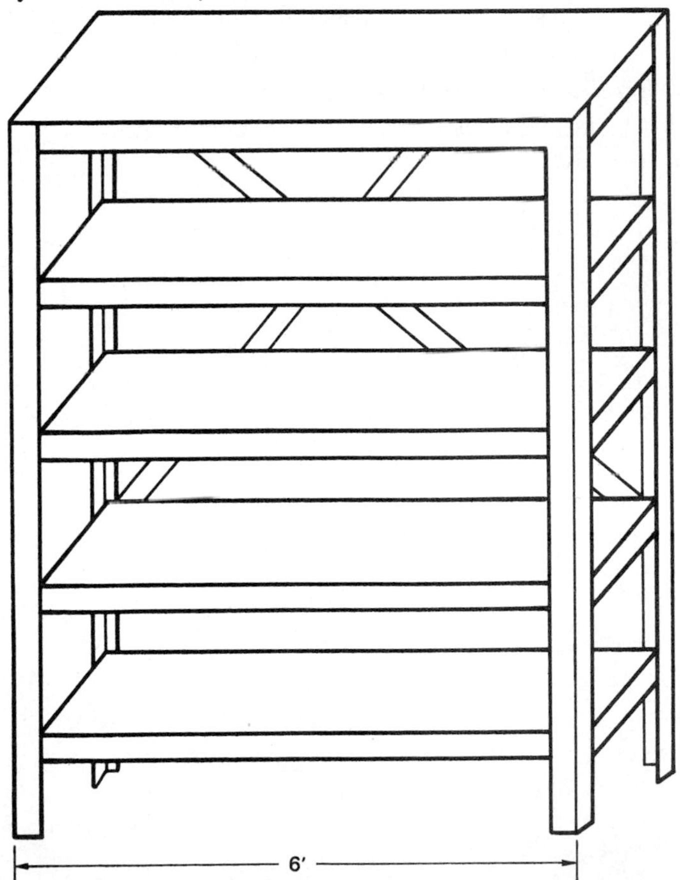

24. An office building is being built. One main supply duct will supply the ventilation flow for the identical offices shown. Each office flow is supposed to be 156 cubic feet/minute (cu ft/min). How many cu ft/min should the main supply plenum be designed to handle? ____________

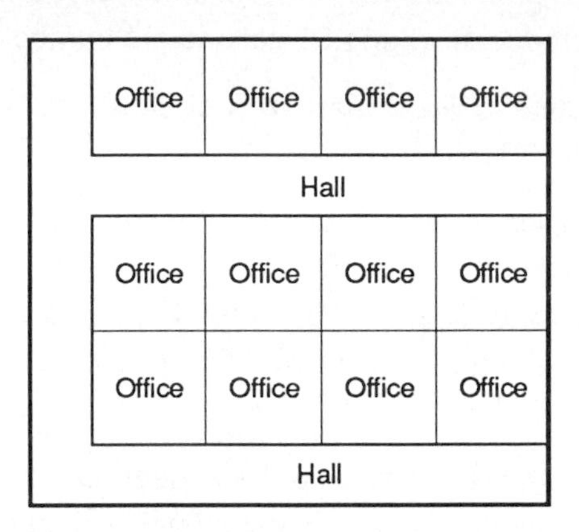

25. One man hour is one man working for one hour. A housing development has 14 buildings. Each building has 6 condominiums in it. Each condominium needs a heating/air conditioning system installed in it. Each system will take 23 man hours to complete the installation. The contractor must plan for how many man hours to complete the job? ____________

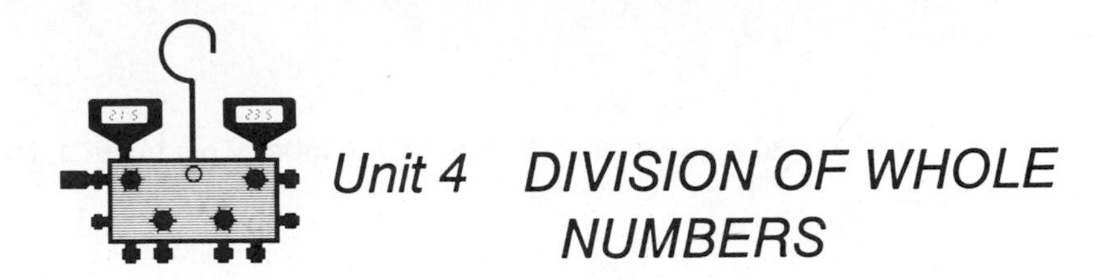

Unit 4 DIVISION OF WHOLE NUMBERS

BASIC PRINCIPLES

- Review and apply the principles of division of whole numbers to the problems in this unit.
- Study division of denominate numbers in section I of the appendix.

Division is probably the most difficult of the mathematical operations. But it can be done correctly time after time if care is taken to do the correct procedure each time. Each of the numbers in a division problem has a special name:

```
        QUOTIENT
DIVISOR)DIVIDEND
```

The problem could also have been given DIVIDEND ÷ DIVISOR.

To perform division, first write the problem down

```
57)194 256
```

Divide the divisor into the first number of the dividend. How many 57's are in 1? In this case the dividend is too small. The divisor will not go into the number. So try the divisor into the first two numbers of the dividend. How many 57's are in 19? Again the 19 is too small. Next try 57 into 194. There are three 57's in 194. Place the 3 in the quotient directly above the 4 in the dividend. There will then be a number in the quotient for each additional number in the dividend.

```
      3
57)194 256
```

Now multiply 57 by 3 and place this number under 194. Subtract this number from 194.

```
      3
57)194 256
   171
   ---
    23
```

Bring down the 2 from the dividend to make 232. Divide 57 into this number. This gives 4. Multiply 57 by 4 and subtract this number from 232.

```
        3 4
57)194 256
   171
    232
    228
      4
```

Bringing down the next number makes 45. Dividing 57 into 45 gives nothing; 45 is too small. However, instead of just bringing down the next number like we did at the beginning of the problem, a 0 is placed in the quotient above the 5. (There must be a number in the quotient for each number in the dividend after the first number has been placed in the quotient.) After this has been done, bring down the next number and continue.

```
        3 408
57)194 256
   171
    232
    228
      456
      456
        0
```

PRACTICAL PROBLEMS

Divide.

1. $3\overline{)69}$
2. $13\overline{)442}$

3. $171\overline{)4\ 446}$
4. $6\overline{)426 \text{ cubic decimeters}}$
5. $42\overline{)14\ 952 \text{ Btu}}$
6. $183\overline{)89\ 304 \text{ amperes}}$
7. 86 ÷ 2 ________
8. 918 ÷ 27 ________
9. 21 239 ÷ 317 ________

10. 392 days ÷ 7 ______

11. 2 268 pounds ÷ 63 ______

12. 245 971 cubic inches ÷ 649 ______

13. A crate contains 4 compressors. The 4 compressors weigh 332 pounds. What is the weight of 1 compressor? ______

14. An apartment complex, with 224 apartments, is being built. Each workday, a heating and air conditioning system can be installed in 7 apartments. How many workdays will it take to install all of the systems? ______

15. A fuel oil tank when filled weighs 1 900 pounds. It stands on 4 legs as shown. What weight must the floor support under each leg; that is, what is the weight each leg supports? ______

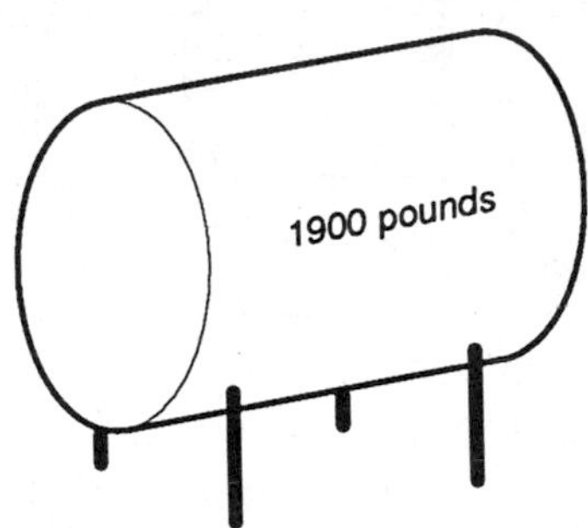

16. New tires are purchased for service trucks. Each service truck is given 4 tires and a spare. If there are 65 new tires, how many trucks will receive new tires? ______

17. This 1 800-square foot attic is to be insulated. One roll of 6-inch thick insulation covers 40 square feet of the attic. How many rolls are needed to insulate the entire attic? ______

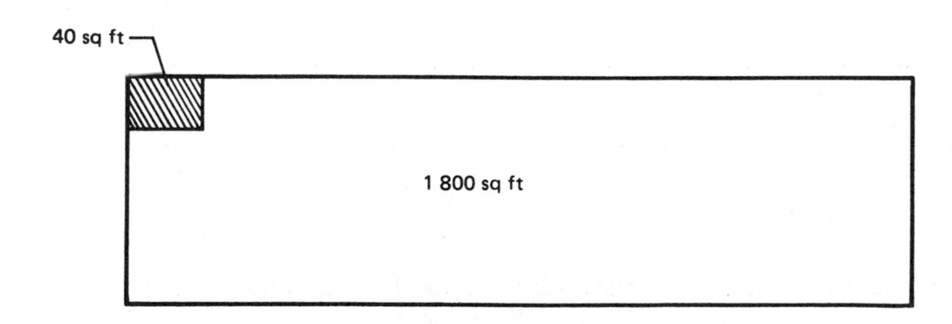

18. The weight of 75 pieces of 3-foot by 8-foot sheet metal is 600 pounds. Find the weight of one sheet. ________

19. To complete the installation of a heating system, 728 hours are needed. Each of the 7 installers works 8 hours a day. How many days will it take to complete the installation? ________

20. One hanger strap is used for the beginning of this duct. The other hanger straps are placed 5 feet apart. How many hanger straps are needed? ________

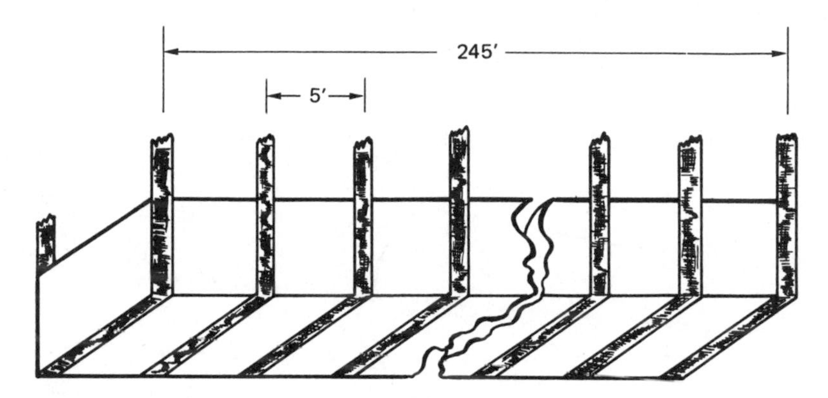

21. A 3-foot by 8-foot piece of sheet metal is used to make 4-inch circular ducts. Each duct is 8 feet long. How many sheets are needed to make 224 feet of the 4-inch circular ducts? ________

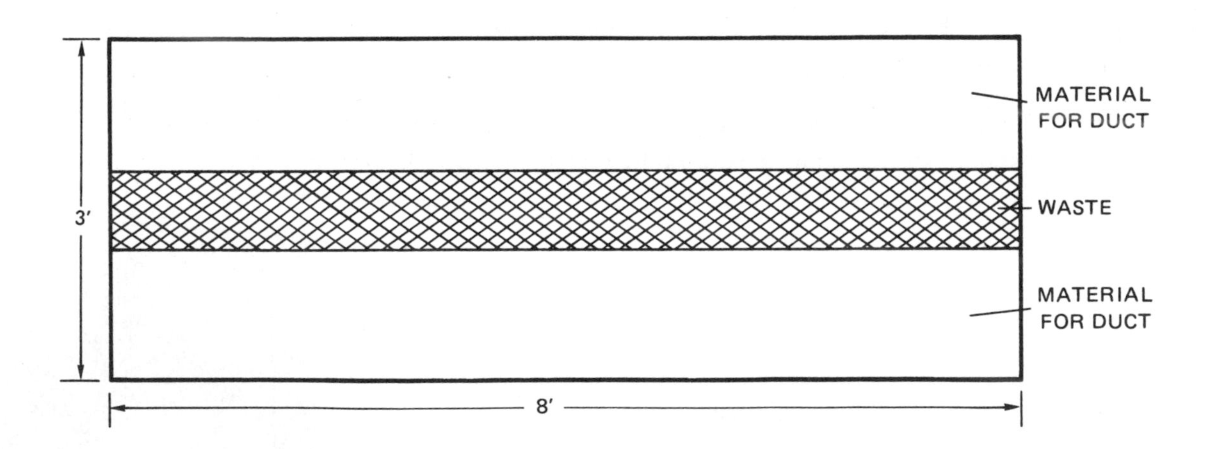

22. In a housing development, 23 identical houses are being built. A hot water heating system is put in each of the houses. A total of 3 496 feet of copper tubing is used. How many feet of tubing are used for each house? ________

23. When 1 cubic foot of natural gas is burned, 1 050 Btu of heat are produced. In one day, a building uses 761 250 Btu of heat. How many cubic feet of gas are burned?

24. Air flows into this room at the rate of 71 cubic feet per minutes. How many minutes will it take to change the air in this room? __________

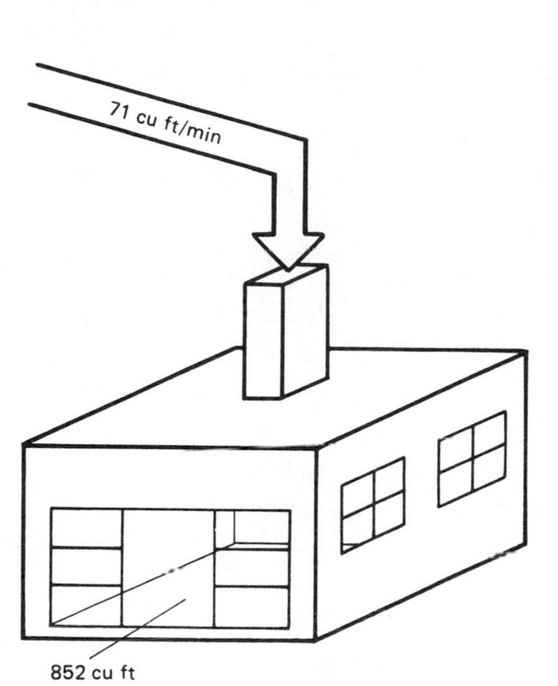

25. A cylinder of R-12 contains 276 ounces of refrigerant. The cylinder is used to recharge auto air conditioners. Each recharge takes 12 ounces. How many car air conditioners can be recharged from this cylinder? __________

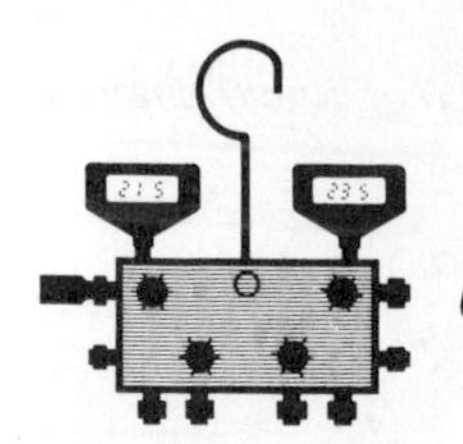

Unit 5 COMBINED OPERATIONS WITH WHOLE NUMBERS

BASIC PRINCIPLES

- Review and apply the principles of addition, subtraction, multiplication, and division of whole numbers to the problems.

PRACTICAL PROBLEMS

1. $\begin{array}{r} 635 \\ 211 \\ 583 \\ +\ 444 \\ \hline \end{array}$

2. $\begin{array}{r} 4\,812 \\ 9\,303 \\ 295 \\ +\ 6\,450 \\ \hline \end{array}$

3. $\begin{array}{r} 643 \\ -\ 218 \\ \hline \end{array}$

4. $\begin{array}{r} 833 \\ -\ 394 \\ \hline \end{array}$

5. $\begin{array}{r} 7\,035 \\ \times\ \ 476 \\ \hline \end{array}$

6. $\begin{array}{r} 2\,884 \\ \times\ 3\,526 \\ \hline \end{array}$

7. $39\overline{)936}$

8. $252\overline{)124\,992}$

9. 137 inches + 5 048 inches + 752 inches ________

10. 265 hours – 88 hours ________

11. 431 seconds × 726 × 89 ________

12. 7 781 kilometers ÷ 31 ________

13. An addition is added to a room. The addition contains 758 cubic feet of space. What is the new volume of the room in cubic feet? ________

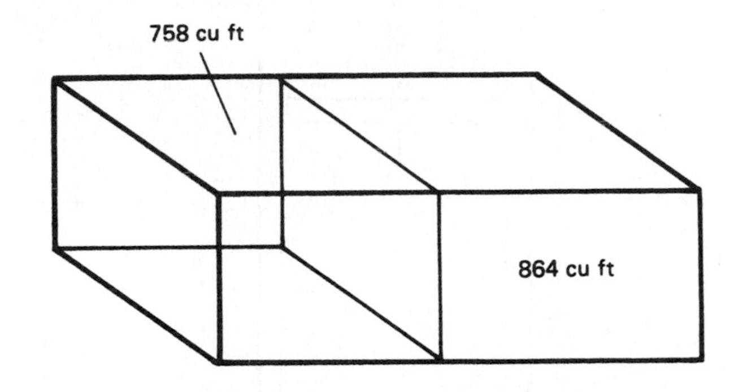

14. The heat loss, in British thermal units per hour (Btu/h), for two different types of walls is given. Find, in British thermal units per hour (Btu/h), the difference in the heat losses. ________

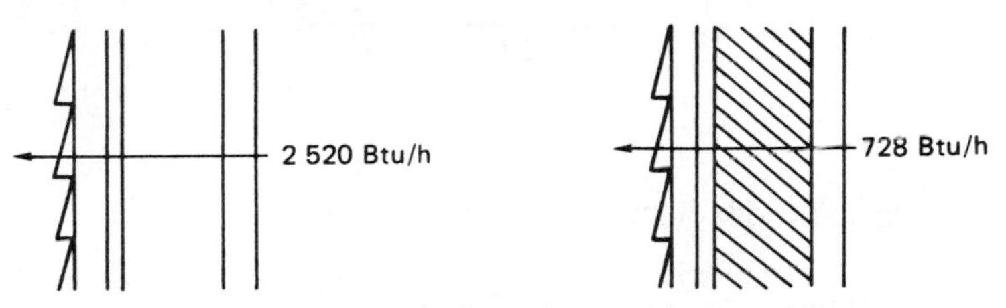

15. A stockroom has 19 boxes of valves. Each box has 4 shut-off valves. What is the total number of shut-off valves in the stockroom? ________

16. A family keeps this record of the number of gallons of fuel oil used to heat their home.

Month	Gallons Used
September	35
October	82
November	144
December	193

Find the total number of gallons used. ________

17. Air flows into a room at the rate of 80 cubic feet per minute. The volume of the room is 960 cubic feet.
    a. How many minutes does it take to change the air in the room? a. ________
    b. How many times will the air be changed in 60 minutes? b. ________

18. The storage capacity of a refrigerator is found by using the inside dimensions of the refrigerator.

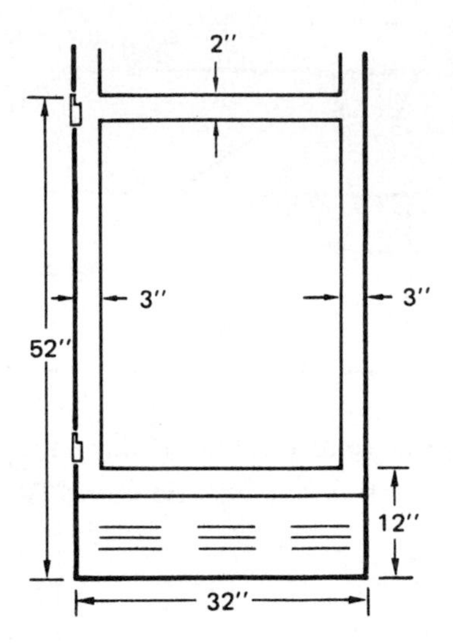

a. Find, in inches, the inside width of this refrigerator. a. ____________

b. Find, in inches, the inside height of this refrigerator. b. ____________

19. At high speed, a blower delivers 2 580 cubic feet of air per minute. This volume is divided equally among 12 ducts. Find, in cubic feet, the amount of air that flows through each duct every minute. ____________

20. A hot water baseboard radiator has 12 fins per inch. How many fins are in 72 inches of the radiator? ____________

21. An office building, with three air conditioning units, has an air conditioning load of 121 000 Btu/h (British thermal units per hour). The first unit handles 41 000 Btu/h and the second handles 42 000 Btu/h. How many British thermal units per hour (Btu/h) does the third unit handle? ____________

22. A forced hot air heater system is used in this mobile home. Find, in cubic feet per minute, the airflow that the central blower must supply. __________

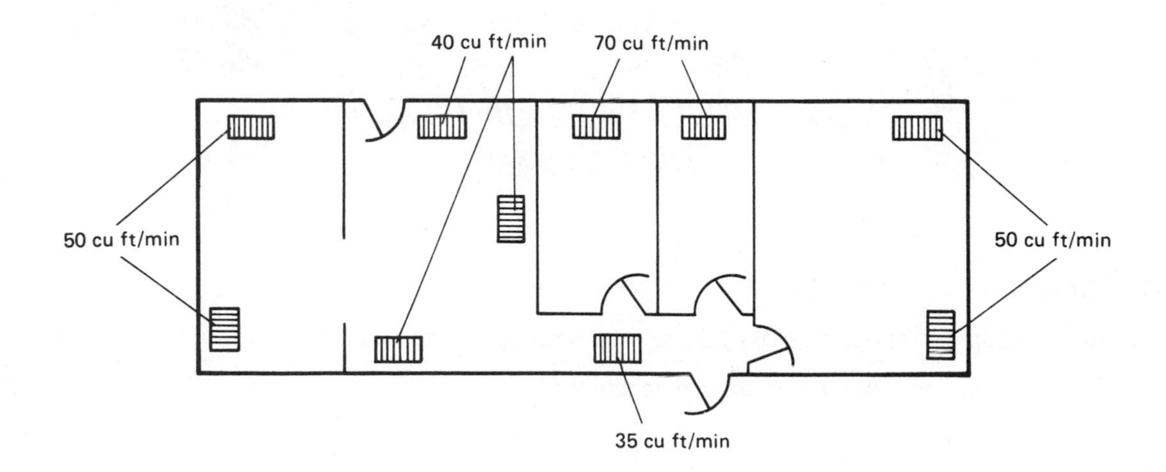

23. An installer finds that it takes 5 minutes to cut out a duct. It then takes 4 minutes to bend the metal and 7 minutes to lap and seal it. If the installer works 50 minutes out of each hour, how many ducts can be made during an 8-hour shift? __________

24. A contractor needs to find the mileage to a job site. The job site is 19 miles from the shop. Two trucks will make one round trip daily. Another truck will make 2 round trips daily. The job will take 24 days to complete. What is the total mileage? __________

25. A one-ton air conditioner is a unit that will remove the amount of heat needed to melt one ton of ice in 24 hours. A ton has 2 000 pounds in it.

    a. If 144 Btu of heat are needed to melt one pound of ice, how many Btu's would a one-ton air conditioner remove in 24 hours? a. __________

    b. How many Btu would a one-ton air conditioner remove in 1 hour? b. __________

Common Fractions

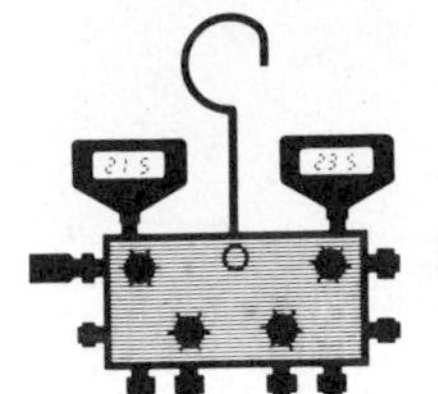

Unit 6 ADDITION OF COMMON FRACTIONS

BASIC PRINCIPLES

- Review and apply the principles of addition of common fractions to the problems in this unit.
- Study addition of denominate numbers in section I of the appendix.

Each fraction is made up of two numbers:

$$\frac{\text{NUMERATOR}}{\text{DENOMINATOR}}$$

When adding fractions, arrange them either in vertical form as whole numbers, or in linear form. In either case, the same procedure must be followed. Fractions cannot be added unless they have the same denominator!

To find the common denominator, take the largest denominator and then make multiples of that number (1 × the number, then 2 × the number, etc.) until you have a number that each of the denominators can be divided evenly into.

$$\frac{3}{4} + \frac{2}{5} + \frac{1}{8}$$

Here, 8 is the largest denominator. Four will divide evenly into 8, however, 5 will not. So take 2 × 8 or 16. Five will not divide evenly into 16, so try 3 × 8 = 24, then 4 × 8 = 32, then 5 × 8 = 40. This is the first multiple of 8 that 5 will divide into evenly. The common denominator, then, is 40.

Equivalent fractions must then be made with 40 as their denominators. An equivalent fraction is one that has the same value as another fraction, but has a different denominator; therefore, it will also have a different numerator. A fraction can be made equivalent by multiplying both numerator and denominator by the same number. In this example, the number you use to multiply is the one needed to make the denominator 40.

$$\frac{3 \times 10}{4 \times 10} + \frac{2 \times 8}{5 \times 8} + \frac{1 \times 5}{8 \times 5} = \frac{30}{40} + \frac{16}{40} + \frac{5}{40}$$

Once the fractions have the common denominator, simply add the numerators.

$$\frac{30 + 16 + 5}{40} = \frac{51}{40}$$

The last step is to see if the fraction can be reduced. This can be done by taking out a whole number if the numerator is larger than the denominator, or dividing both numerator and denominator by the same number. This gives the final answer.

$$\frac{51}{40} = \frac{40}{40} + \frac{11}{40} = 1\frac{11}{40}$$

With mixed numbers (a whole number and a fraction), work with the fractions and then with the whole numbers.

$$3\frac{3}{4} + 5\frac{5}{6} =$$

$$\frac{3}{4} + \frac{5}{6} = \frac{3 \times 3}{4 \times 3} + \frac{5 \times 2}{6 \times 2}$$

$$= \frac{9}{12} + \frac{10}{12}$$

$$= \frac{9+10}{12} = \frac{19}{12}$$

$$= 1\frac{7}{12}$$

$$3 + 5 + 1\frac{7}{12} = 9\frac{7}{12}$$

Keep in mind the following guidelines when adding common fractions:

- Find the lowest common denominator
- Make equivalent fractions with lowest common denominators
- Reduce the answer to lowest terms.

PRACTICAL PROBLEMS

Add.

1. $\frac{3}{5}$, $\frac{1}{5}$, $\frac{4}{5}$, $+\frac{3}{5}$

2. $\frac{1}{6}$, $\frac{4}{9}$, $\frac{2}{3}$, $+\frac{1}{2}$

3. $\frac{4}{5}$, $\frac{3}{10}$, $\frac{6}{25}$, $+\frac{7}{50}$

4. $\frac{5}{18}$, $\frac{3}{4}$, $\frac{1}{3}$, $\frac{5}{6}$, $+\frac{4}{9}$

5. $22\frac{5}{16}$, $7\frac{3}{4}$, $\frac{1}{8}$, $15\frac{1}{2}$

6. 3/4 inch + 1/8 inch + 5/64 inch + 3/8 inch + 7/32 inch __________

7. 9/16 inch + 5/8 inch + 1/2 inch + 1/32 inch + 1/4 inch __________

8. $2\frac{1}{3}$ pounds + $15\frac{4}{5}$ pounds + 7 pounds + $6\frac{3}{4}$ pounds __________

9. Find, in inches, the width of the galvanized sheet steel needed to make this duct. __________

10. What is the center-to-center distance of the two bolt holes on this conduit pipe strap? __________

11. This filter-drier is to be added to a refrigerant circuit. What is the length this adds to the tubing? ____________

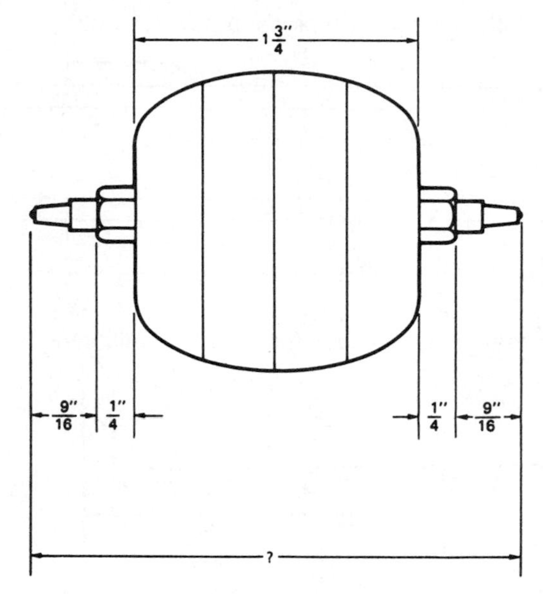

12. A fuel line must pass through this wall. Find, in inches, the length of the hole through the wall. ____________

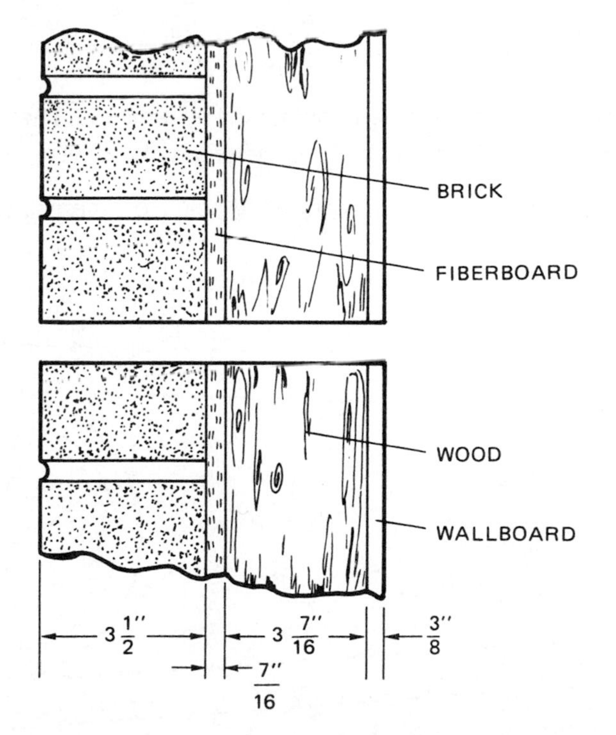

13. A defective section of soft copper tubing must be replaced. A 21 5/16-inch section is cut out. The tube to replace the defective section will overlap 3/8 inch at each end. How many inches long is the piece of new tubing? __________

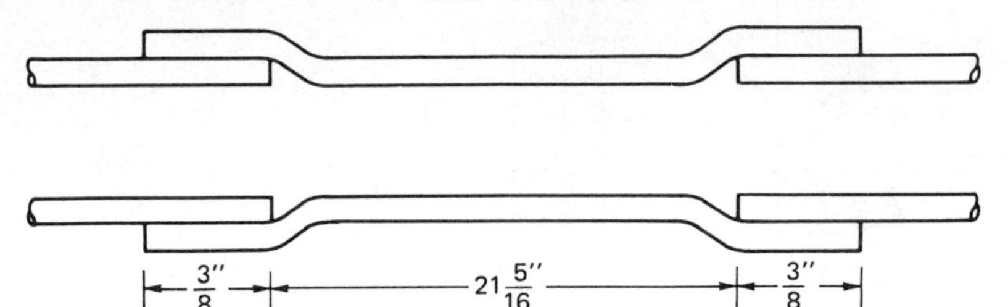

14. What is the length of window seal gasket needed for this air conditioning unit? __________

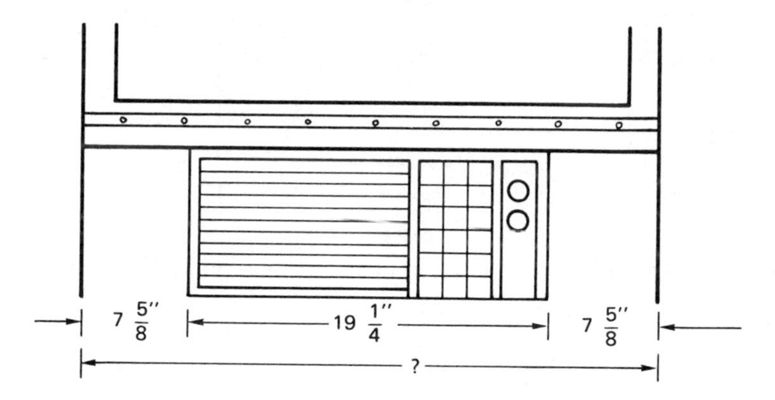

15. A worn magnetic gasket around a refrigerator door must be replaced. The door is 30 1/4 inches wide and 40 1/8 inches high. Find the length of gasket needed. __________

16. A cooling fan is connected to the driving motor by a V-belt. How many inches long is the V-belt? __________

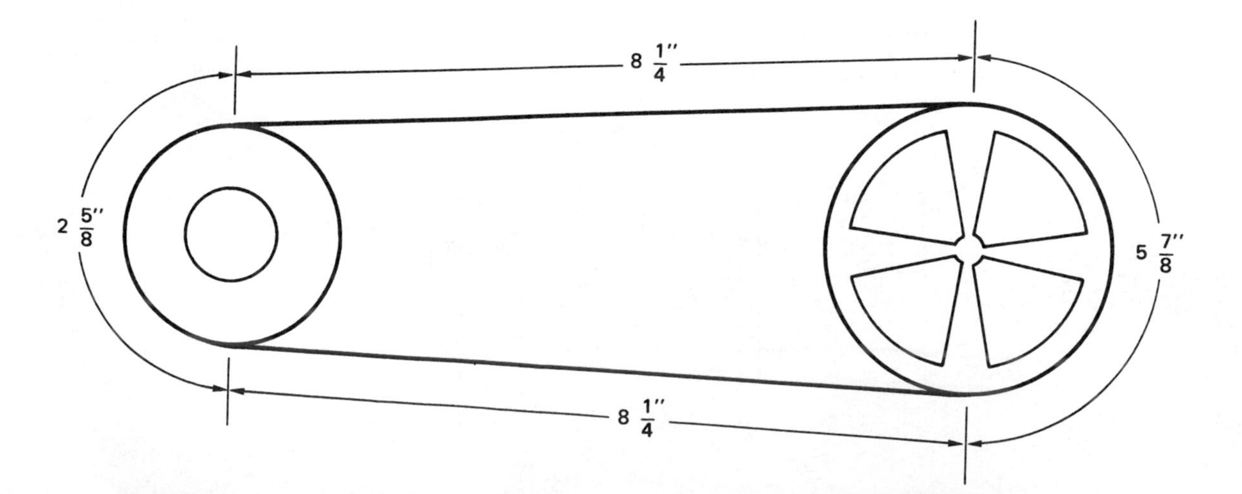

17. Find, in feet, the distance the air flows through this duct. __________

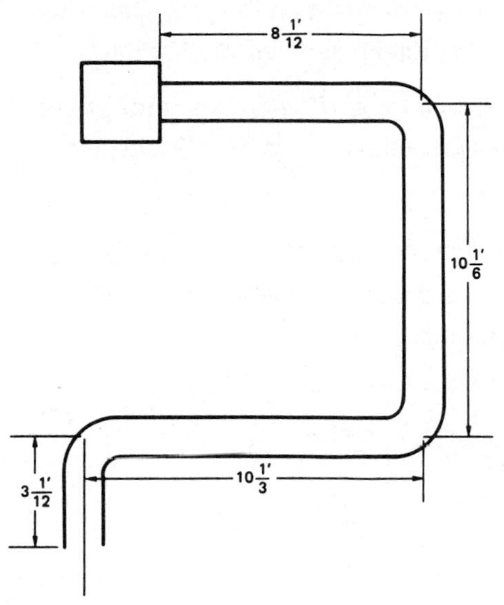

18. A conduit for the power wires of an air conditioning unit is put together. How long is the conduit? __________

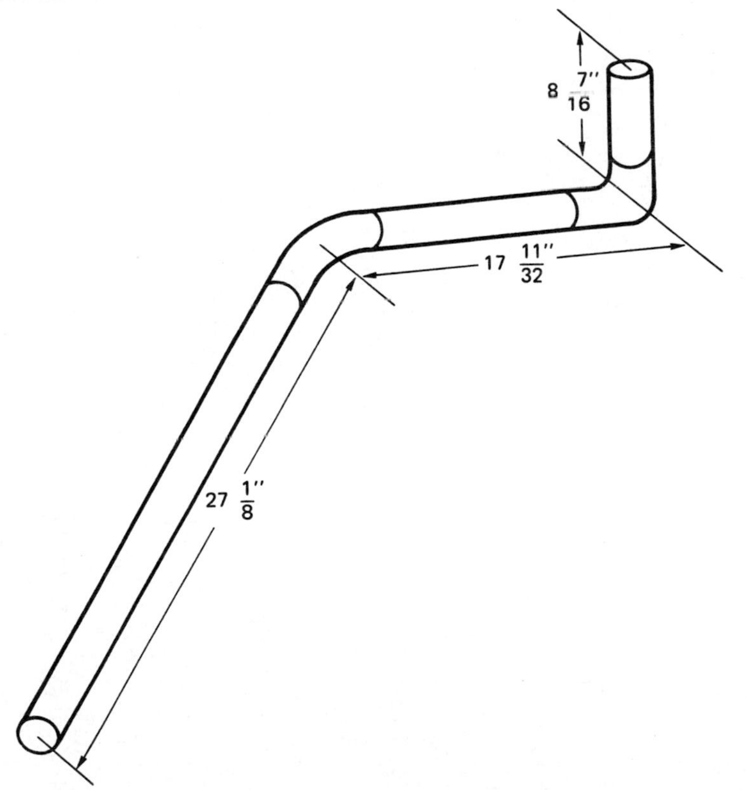

19. A repairer had to work on a single job on 3 different days. Monday 6 3/4 hours were spent on the job; Tuesday 4 1/2 hours were spent on the job; Thursday 6 hours were spent on the job. What was the total time spent on the job? __________

20. A repairer needs to drill a hole through a wall for a 1/2-inch fuel line. There should be a 1/32-inch clearance on each side of the tube. What should be the diameter of the drill? __________

21. A piece of ducting must be custom made. It must fill a gap 25 7/16-inches long. One end of the duct must also fit into the next piece of ducting 2 3/8 inches. What is the total length this piece of ducting must be? __________

Unit 7 SUBTRACTION OF COMMON FRACTIONS

BASIC PRINCIPLES

- Review and apply the principles of subtraction of common fractions to the problems in this unit.
- Study subtraction of denominate numbers in section 1 of the appendix.

Subtracting fractions is almost like addition. You find the common denominator and make equivalent fractions. Subtract the numerators. A major difference between addition and subtraction is in the case of mixed numbers. There are times when you may have to borrow from the units and make the number a fraction.

$$5\frac{1}{3} - 2\frac{3}{5} = 5\frac{1 \times 5}{3 \times 5} - 2\frac{3 \times 3}{5 \times 3}$$

$$= 5\frac{5}{15} - 2\frac{9}{15}$$

$$= 4\frac{5 + 15}{15} - 2\frac{9}{15}$$

$$= 4\frac{20}{15} - 2\frac{9}{15}$$

$$= 2\frac{20 - 9}{15} = 2\frac{11}{15}$$

Use the following guidelines when subtracting common fractions:

- Find the lowest common denominator.
- Make equivalent fractions with lowest common denominators.
- Reduce the answer to lowest terms.
- You may have to borrow to complete the subtraction.

PRACTICAL PROBLEMS

Subtract.

1. $\frac{7}{16} - \frac{5}{16}$
2. $\frac{2}{3} - \frac{4}{25}$
3. $3\frac{4}{5} - 1\frac{1}{3}$
4. $2\frac{2}{5} - 1\frac{7}{8}$
5. $4\frac{1}{16} - \frac{4}{5}$
6. 1/4 inch – 1/16 inch
7. 9/32 inch – 1/8 inch
8. $9\frac{7}{8}$ inches – $7\frac{5}{8}$ inches
9. $15\frac{1}{4}$ inches – $9\frac{7}{8}$ inches
10. A piece of PVC tubing, 3 3/8 feet long, is needed for a drain of an air conditioning unit. The piece is cut from an 8-foot long coil of tubing. How much tubing is left?
11. How many inches does this window air conditioner extend inside the window?

12. This air conditioning condenser unit is mounted on the top of a flat-roofed building. Find, in inches, dimension A. ____________

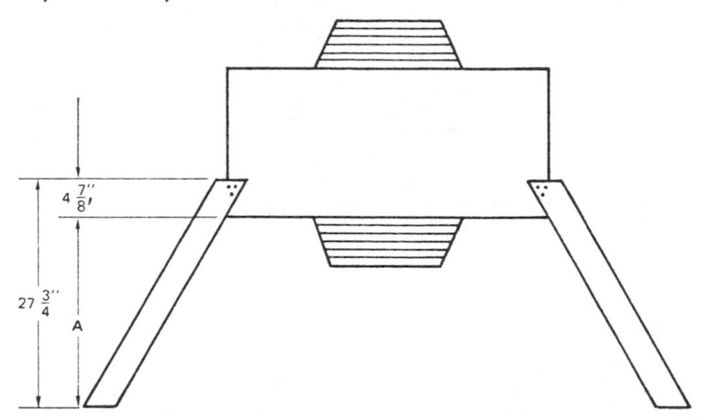

13. In one day, a repairer works 9 1/2 hours and finishes two jobs. It takes 3 3/4 hours to finish the first job. How long does it take to finish the second job? ____________

14. Find, in inches, the drop in height for this fuel line. ____________

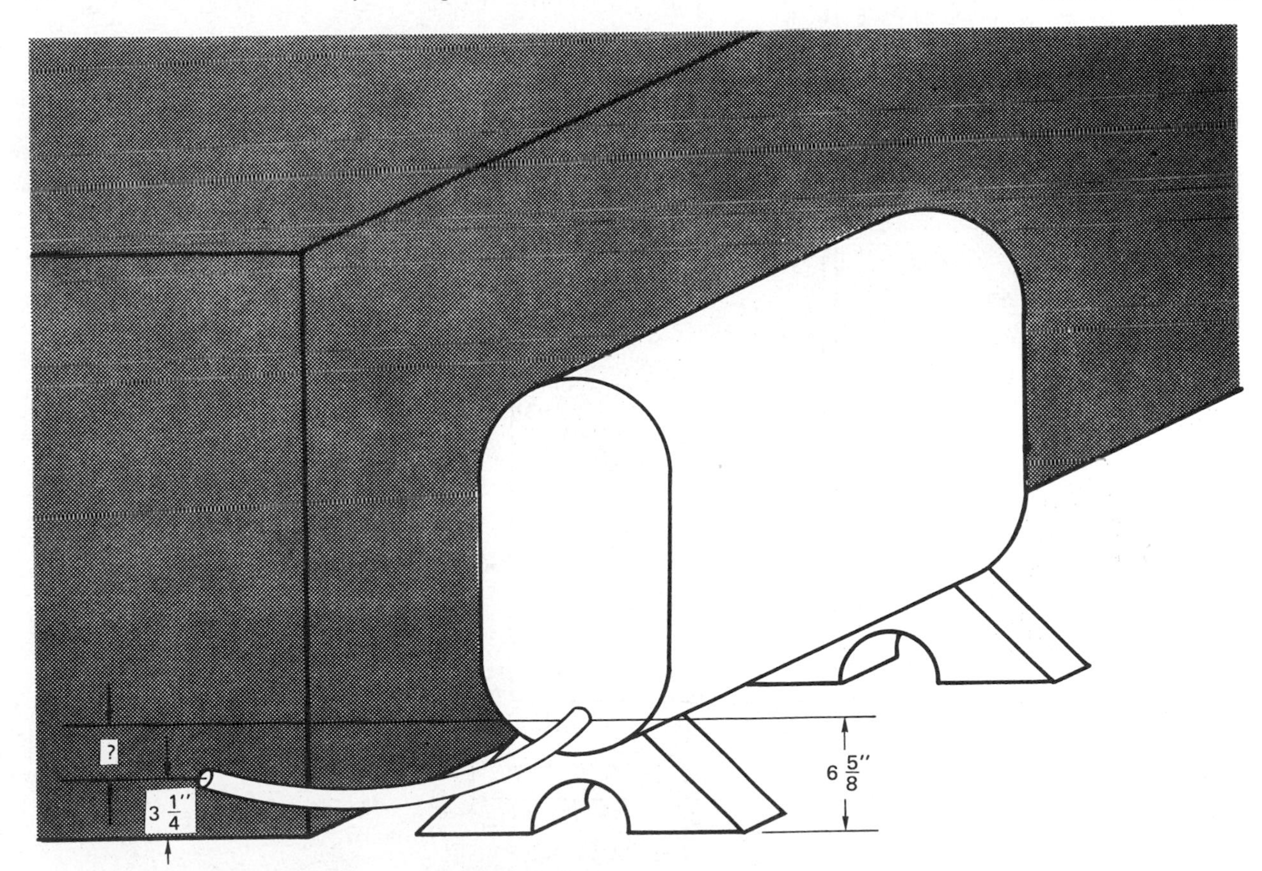

15. A spray nozzle humidifier is installed in this air duct.

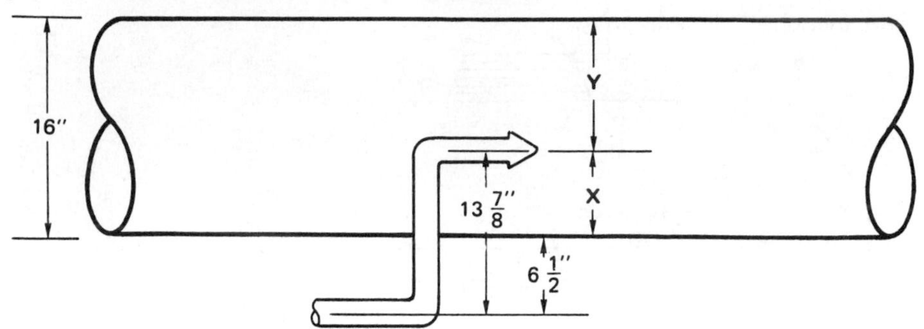

a. Find, in inches, dimension X. a. ______________

b. Find, in inches, dimension Y. b. ______________

16. Determine dimension A on this refrigerator. ______________

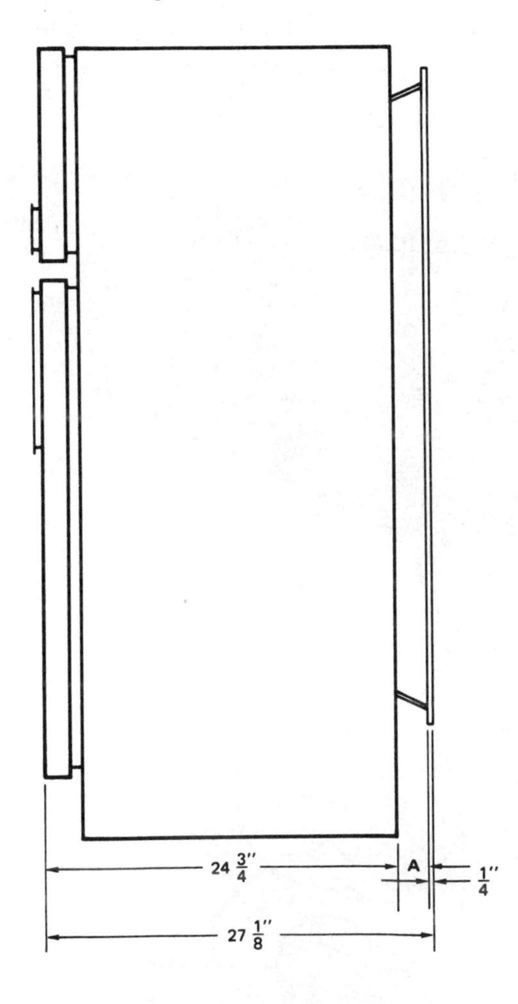

17. This vent cover is fitted into a duct.

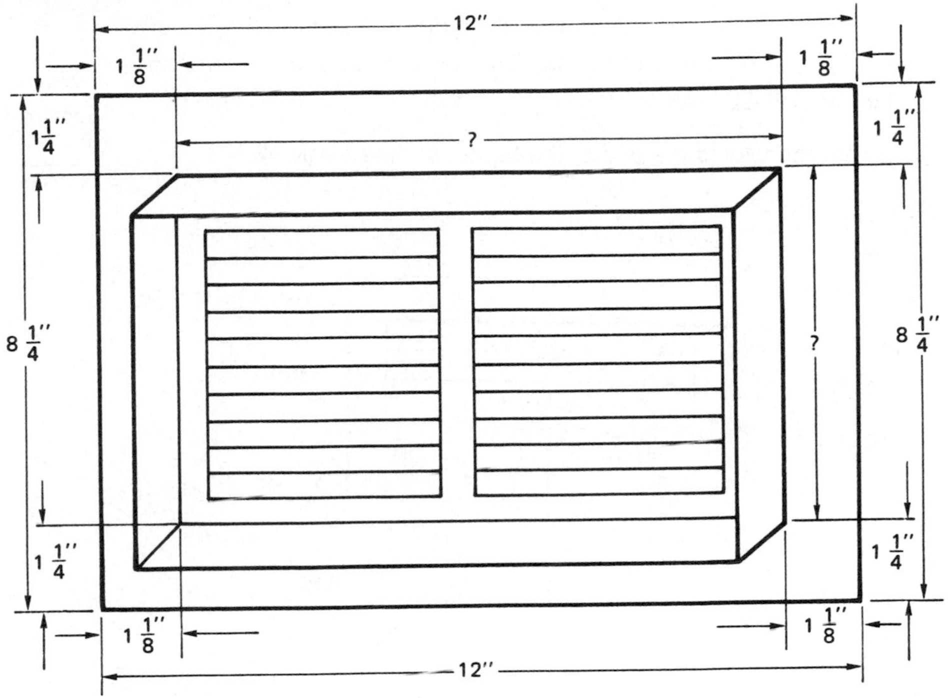

a. How many inches high must the duct be? a. ______________

b. How many inches wide must the duct be? b. ______________

18. What is the length of the finned section of pipe in this baseboard hot water heating system? ______________

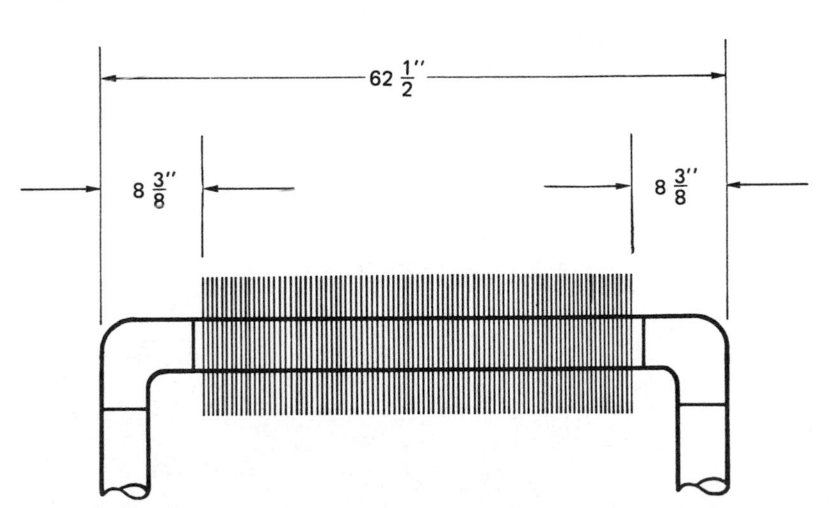

19. Five air conditioning units are checked and recharged with refrigerant R-22. A full cylinder of R-22 contains 25 pounds of refrigerant. From a full cylinder 3 1/2 pounds, 2 1/3 pounds, 4 2/3 pounds, 2 4/5 pounds, and 5 1/6 pounds are taken. How much R-22 is left? __________

20. How far away from the wall is the end of the heater pictured below? __________

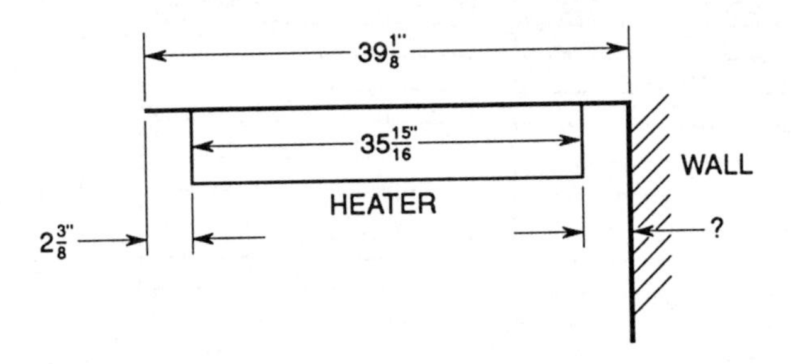

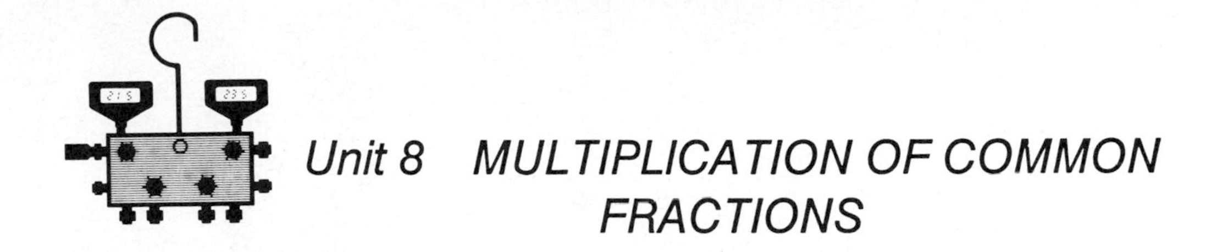

Unit 8 MULTIPLICATION OF COMMON FRACTIONS

BASIC PRINCIPLES

- Review and apply the principles of multiplication of common fractions to the problems in this unit.
- Study multiplication of denominate numbers in section I of the appendix.

Multiplying fractions is easier than adding or subtracting them. It is best to put the fractions in a linear form. Then it is simply a case of multiplying the numerators and also multiplying the denominators.

$$\frac{2}{3} \times \frac{4}{5} = \frac{2 \times 4}{3 \times 5} = \frac{8}{15}$$

Once again the last thing to do is to check if the answer can be reduced. The reducing can take place in an interim step: before multiplying see if there is any number that can divide into one numerator and one denominator. In the example below, the factor is 3.

$$\frac{2}{\underset{1}{\not{3}}} \times \frac{\overset{3}{\not{9}}}{11} = \frac{2 \times 3}{1 \times 11} = \frac{6}{11}$$

When multiplying, do not multiply mixed numbers. Change all mixed numbers to fractions before multiplying. You will probably have to change them back to mixed numbers when you are done.

$$1\frac{1}{3} \times 2\frac{5}{6} = \frac{4}{3} \times \frac{17}{6}$$

$$= \frac{2 \times 17}{3 \times 3} = \frac{34}{9}$$

$$= 3\frac{7}{9}$$

Follow these guidelines when multiplying common fractions:

- Remember when multiplying fractions, you do not need to find the lowest common denominator.
- When cancelling, one of the numerators and one of the denominators must be divided by the same number.
- You do not have to cancel; cancelling just makes the problem easier because you work with smaller numbers.
- You may cancel more than once.
- Reduce the answer to lowest terms.

PRACTICAL PROBLEMS

1. $\frac{1}{4} \times \frac{3}{4}$

2. $\frac{1}{3} \times \frac{3}{7}$

3. $\frac{2}{3} \times \frac{4}{5}$

4. $\frac{2}{7} \times \frac{49}{50}$

5. $4\frac{1}{5} \times \frac{2}{3}$

6. $2\frac{1}{4} \times 1\frac{2}{5}$

7. $1\frac{1}{35} \times 1\frac{3}{4}$

8. 1/2 inch × 1/4 ______

9. 1 1/6 feet × 1/3 ______

10. 3 1/3 feet × 2 1/20 ______

11. 3/5 × 1 1/4 inches × 3 1/6 ______

12. The minimum safe bending radius of a certain polyethylene tube is the outside diameter times 3 1/2. The tube has an outside diameter of 3/8 inch. What is the minimum bending radius? ______

13. A house is being built. The contractor states that the cost to install electric baseboard heat will be $440. He also states that a forced air, heat pump system will be 2 1/4 times as much. How much will the heat pump system cost? ______

14. What is the total distance between the centers of the first and last gas burners? ______

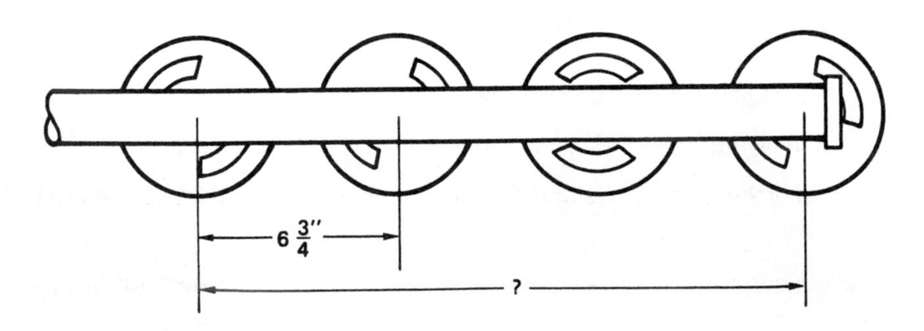

15. A supply plenum is fastened to the furnace with equally spaced screws. Find, in inches, dimension X. __________

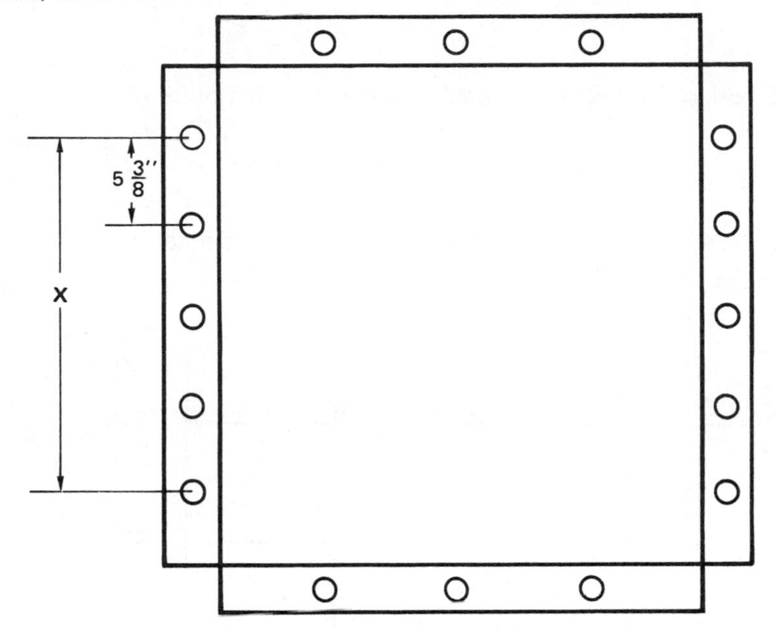

16. A fuel tank holds 281 1/2 gallons. One gallon of #2 fuel weighs 7 1/10 pounds. How many pounds of fuel are in a full tank? __________

17. Support straps are used for this duct. Each support strap is made from two pieces of metal each 21 3/8 inches long. How many inches of metal are needed for the straps? __________

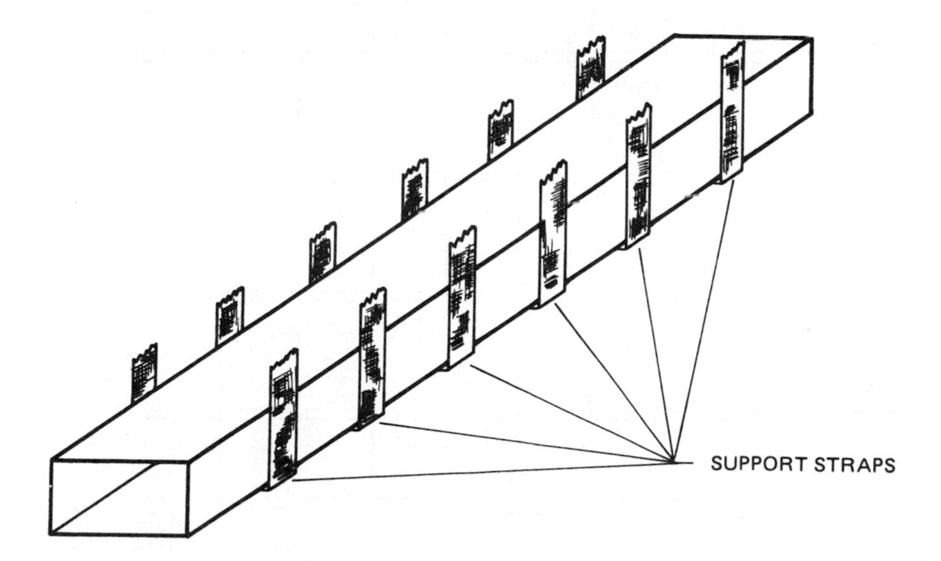

18. Each hour an oil burner runs, the nozzle sprays 3/4 gallon of fuel. If the burner runs for 2 1/2 hours, how much fuel is sprayed into the burner? __________

19. Washers which fit a 3/16-inch bolt cost 5/6¢ each. What is the cost of 36 (3 dozen) washers? __________

20. This condenser coil is made from tubing which has a 1/4-inch outside diameter. How many inches of tubing are needed for the coil? __________

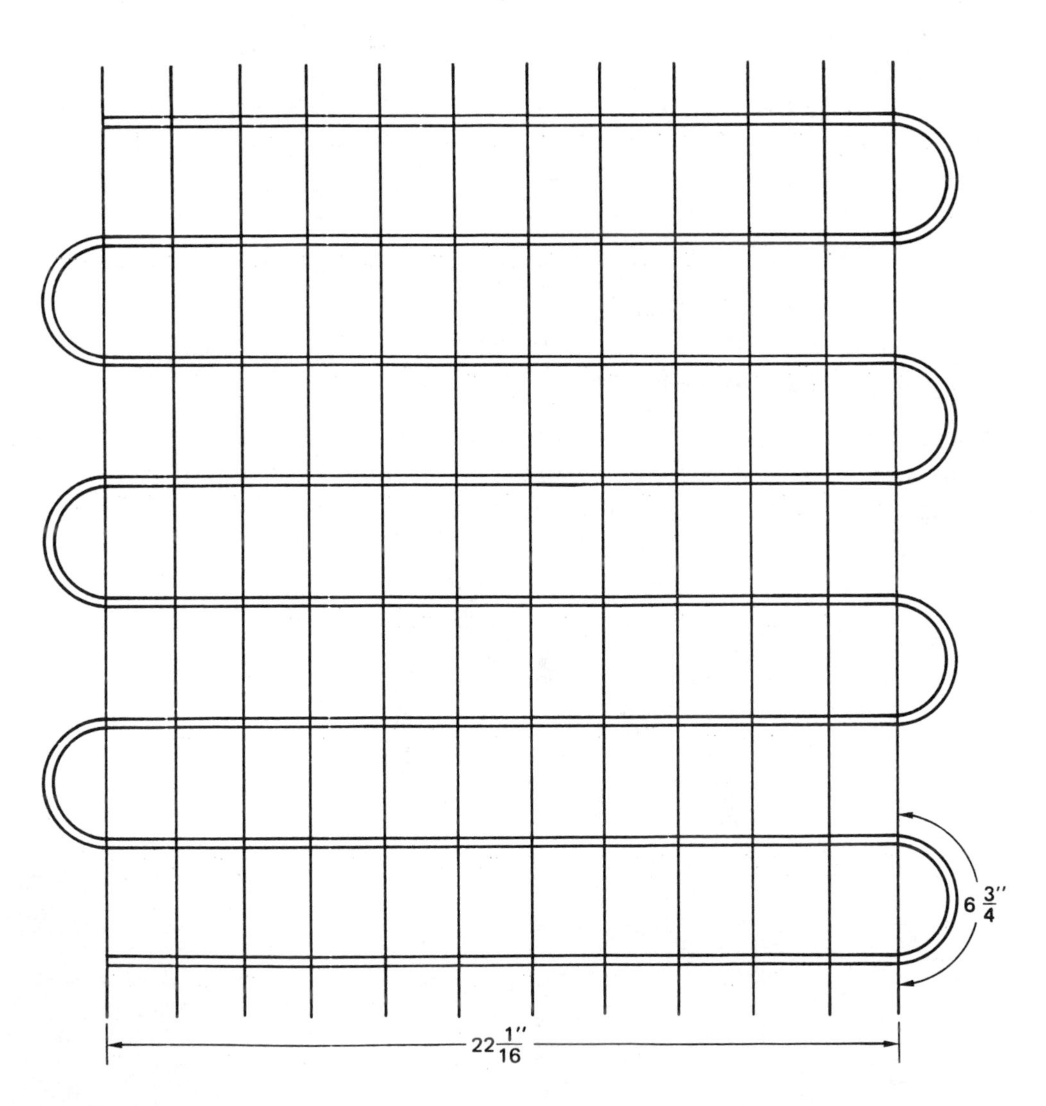

21. A repairer spends 9 hours at work. Repairing a refrigerator takes 1/2 of the time and repairing a window air conditioner takes 1/3 of the time. The repairer spends 1/18 of the time ordering parts and 1/9 of the time eating lunch. Find the number of hours the repairer spends in each activity.

    a. repairing refrigerator — a. ____________

    b. repairing air conditioner — b. ____________

    c. ordering parts — c. ____________

    d. eating lunch — d. ____________

22. When installing a hot water heating system, these pieces of 1/2-inch tubing are used: five pieces, each 8 1/4 inches long; four pieces, each 24 1/8 inches long; five pieces, each 5 7/16 inches long. What length of 1/2-inch tubing is needed for this project? ____________

Unit 9 DIVISION OF COMMON FRACTIONS

BASIC PRINCIPLES

- Review and apply the principles of division of common fractions to the problems in this unit.
- Study division of denominate numbers in section I of the appendix.

Division of fractions is very similar to multiplication of fractions with one exception. Take the divisor (the fraction after the ÷ sign) and invert it (make the numerator the denominator and the denominator the numerator). Then change the ÷ sign to a × sign and multiply the two fractions.

$$1\frac{1}{3} \div 1\frac{1}{2} = \frac{4}{3} \div \frac{3}{2}$$
$$= \frac{4}{3} \times \frac{2}{3}$$
$$= \frac{8}{9}$$

Use the following guidelines when dividing common fractions:

- Invert the fraction after the ÷ sign and change the ÷ sign to a × sign.
- After inverting, treat the problem as a multiplication problem.
- Do not cancel before inverting.

PRACTICAL PROBLEMS

Divide.

1. 2/5 ÷ 7/9 __________
2. 3 1/5 ÷ 4/7 __________
3. 3 ÷ 1 2/3 __________
4. 1 7/25 ÷ 2/5 __________
5. 5 4/9 ÷ 8 2/3 __________
6. 6 2/9 ÷ 1 17/18 __________
7. 1/8 inch divided by 2 __________
8. 3/4 hour divided by 3 __________
9. 3/7 week divided by 1/4 __________
10. 6 2/3 yards divided by 2/3 yard __________

11. Air duct straps are made from a 1-inch wide strip of metal. The metal is 246 3/4 inches long. Each strap is 17 5/8 inches long. How many straps can be made? __________

12. Air conditioning systems are charged with refrigerant R-12. The cylinder of refrigerant contains 25 pounds. Each system uses 1 3/4 pounds of R-12. How many complete systems can be charged from the 25-pound cylinder? __________

13. A section of a condenser measures 4 1/4 inches and has 51 fins. Some of the fins are bent and must be straightened with a fin comb. The fin comb must have the same number of fins per inch as the condenser. How many fins per inch does the fin comb have? __________

14. Strips of metal are cut from a 3-foot by 8-foot sheet. Each strip is 3 feet long and 19 3/4 inches wide. Find the number of complete strips that can be made from this sheet. __________

15. Single wall smoke pipe is used to connect the oil burner to the chimney flue. The distance from the burner to the vent is 157 1/2 inches. If 1 1/2 inches are allowed on each pipe for fitting, how many lengths of pipe are needed? __________

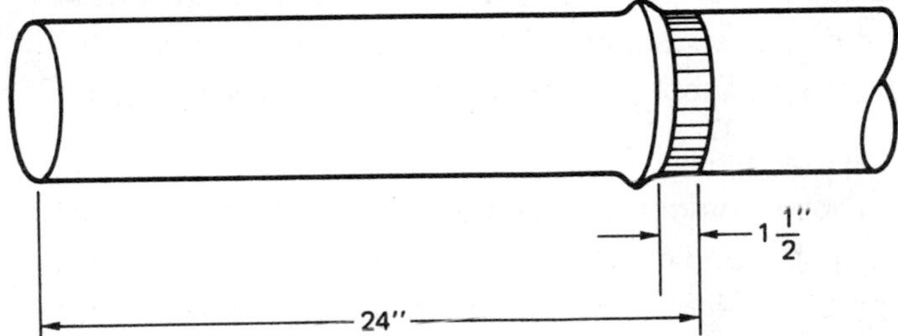

16. Two filler boards are used to install this window air conditioning casing. If the two boards are equal in width, find each width. __________

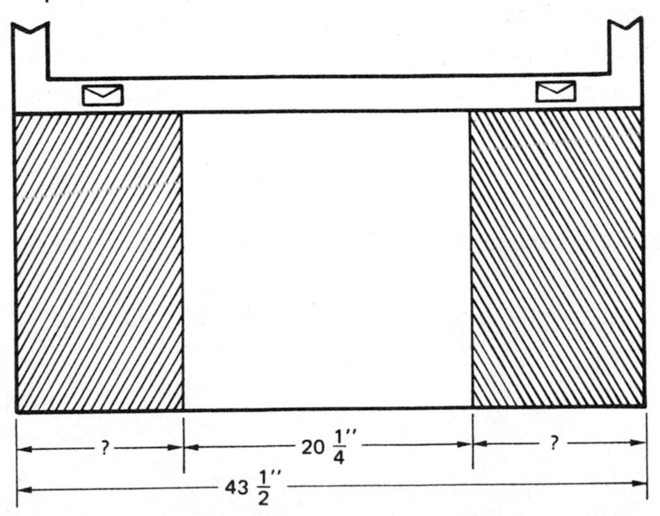

17. This ceiling register has equally spaced openings.

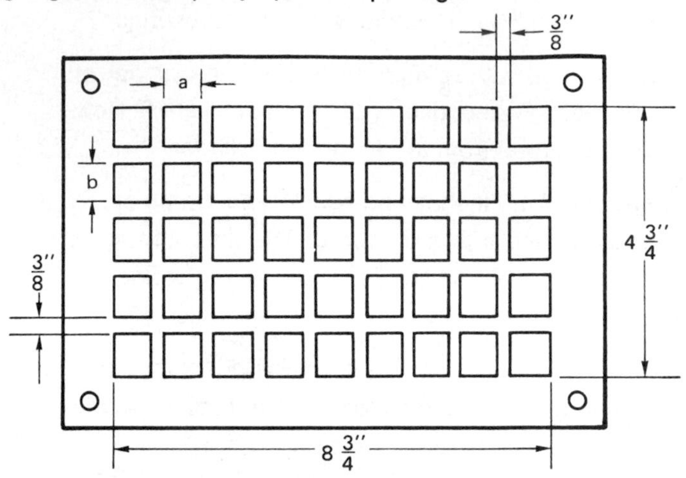

a. Find, in inches, the width of each opening. a. ____________

b. Find, in inches, the height of each opening. b. ____________

18. A certain house has floor joists every 1 1/3 feet. The heating duct for the house runs under the floor. The duct is supported by straps that are attached to the joists. The distance between the support straps is 4 feet. How many joists does this represent? ____________

19. Duct tape is used to seal joints in duct insulation. A roll of duct tape contains 360 inches of tape. A 10-inch round duct with a 1-inch layer of insulation on it will need a piece of tape 37 3/4-inches long to cover the seam and overlap itself. How many seams will 1 roll of duct tape cover? ____________

20. A ventilating system is to be installed in a building. It will require 400 hours of work to put it in. Each workman takes a total of 1/2 of an hour to get out his tools in the morning and put them away at night. He also takes a total of 1/4 of an hour to get to his workplace and back again each day. He is also given a 1/4-hour break in the morning and another in the afternoon.

a. For an 8-hour workday, how much time is actually spent installing the ventilating system? a. ____________

b. If 5 men are in the work crew, how many hours of work are done each day? b. ____________

c. How many days will be needed to complete the job with a 5-man crew? c. ____________

Unit 10 COMBINED OPERATIONS WITH COMMON FRACTIONS

BASIC PRINCIPLES

- Review and apply the principles of addition, subtraction, multiplication, and division of common fractions to these problems.

PRACTICAL PROBLEMS

1. $2\frac{1}{5} + 3\frac{5}{9}$

2. $\frac{11}{12} + 2\frac{3}{4}$

3. $5\frac{2}{3} + 7\frac{7}{8} + 2\frac{5}{6}$

4. $4\frac{5}{6} - 2\frac{1}{2}$

5. $3\frac{6}{7} - 2\frac{1}{3}$

6. $4\frac{1}{5} - 3\frac{6}{7}$

7. $\frac{5}{6} \times \frac{1}{4}$

8. $3\frac{1}{3} \times 2\frac{1}{11}$

9. $5\frac{5}{6} \times 2\frac{2}{7}$

10. 2/3 ÷ 5/7 ________

11. 3/4 ÷ 1/5 ________

12. 2 4/9 ÷ 2 1/5 ________

13. 1/6 yard + 3/8 yard ________

14. 7/8 inch – 5/32 inch ________

15. 15/16 pound × 4/5 ________

16. 4 1/6 yards ÷ 3 1/3 ________

17. An oil burner is installed in a basement. The floor of the basement slopes and the oil burner must be leveled. What is the height from the floor to the top of the burner after leveling? __________

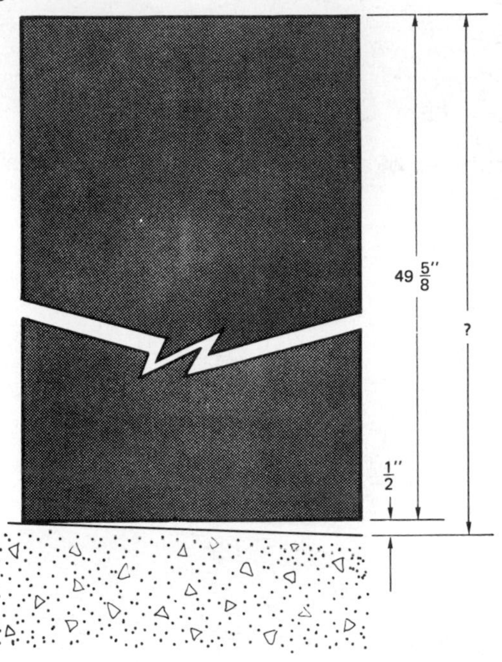

18. An installer uses *#10* wire for a repair job. Before starting the job, the roll has 78 1/2 feet of wire on it. The installer uses 1/3 of the roll. How many feet of wire does the installer use? __________

19. These 5 air conditioning condensing units are mounted next to each other on a roof. Each unit is equal in size. What is the width of each unit in inches? __________

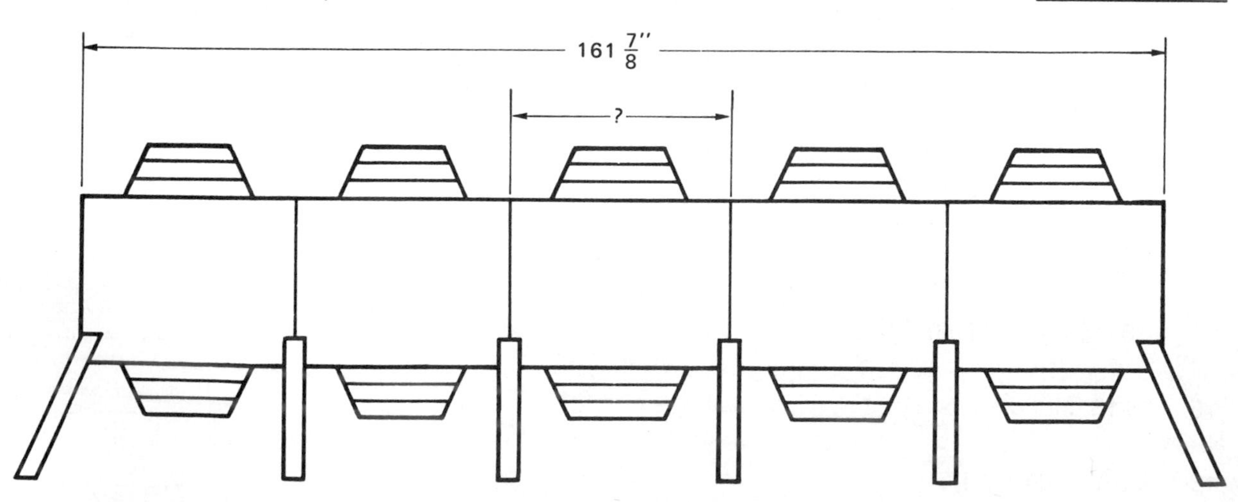

20. The nozzle and electrodes of an oil burner extend from the end of the burner gun. Find, in inches, dimension X. __________

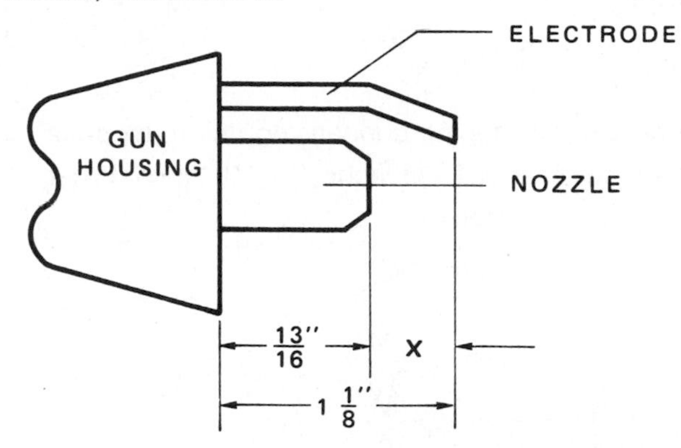

21. What is the total length of this furnace and air conditioning system? __________

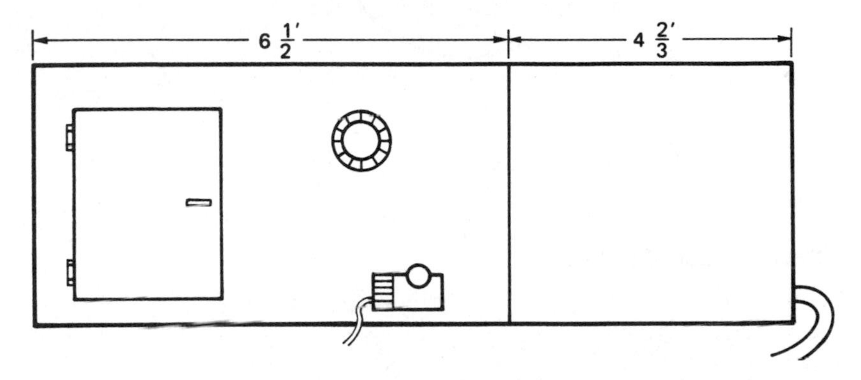

22. Washers are placed on a levelling screw to give it support. The washers are 18 gauge and are 3/64 inch thick. How many washers are needed to fill the space? __________

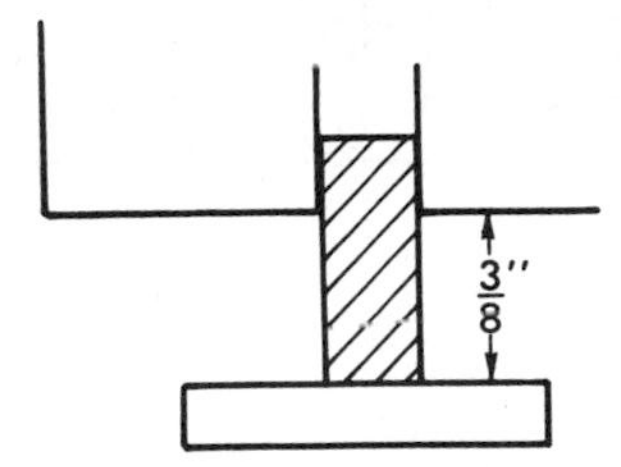

23. At the start of a workday, a cylinder of refrigerant R-22 has 22 1/2 pounds in it. A repairer works on 4 air conditoning units and uses all of the refrigerant. On the first unit 1/4 of the refrigerant is used and and 1/12 of the refrigerant is used on the second unit. Then 1/6 of the refrigerant is used on the third unit and 1/2 on the fourth. Find the number of pounds of refrigerant R-22 used on each unit.

a. Unit #1
b. Unit #2
c. Unit #3
d. Unit #4

a. ___________
b. ___________
c. ___________
d. ___________

24. The tubes going to and from an air conditioning condensing unit must pass through a wall. The tubes have diameters of 1 7/8 inches and 3/4 inch. What is the smallest diameter hole that can be used? ___________

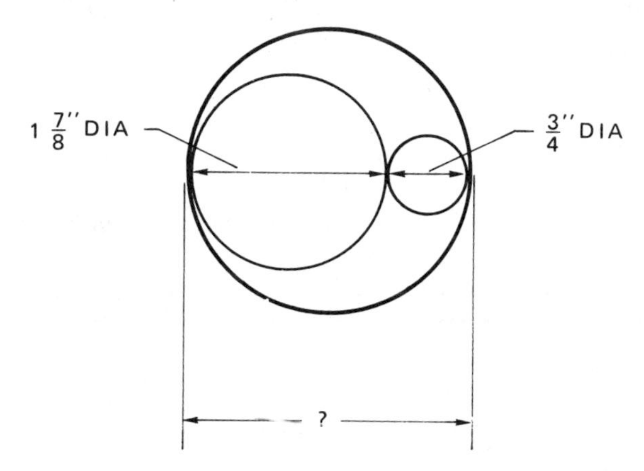

25. Find, in inches, the diameter of this duct. ___________

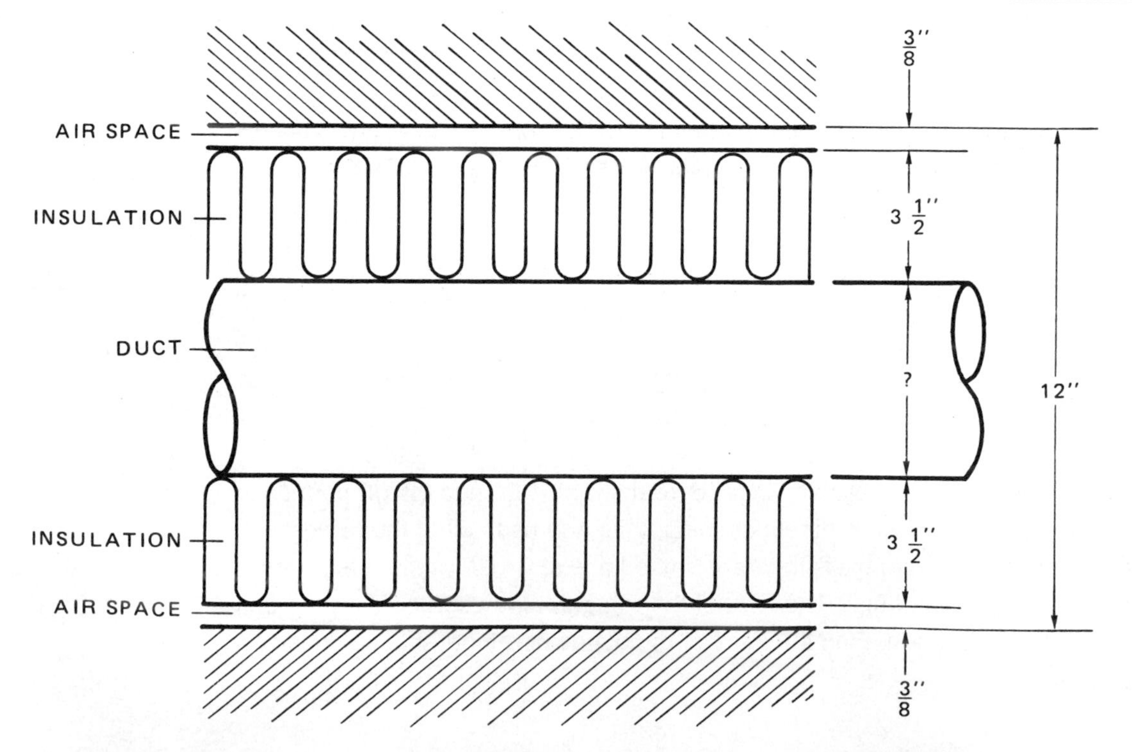

26. A circular duct has an outside diameter of 7 1/2 inches. Insulation, 1 1/8 inches thick, is wrapped around the duct. What is the diameter of the insulated duct? __________

27. A piece of *12* gauge sheet metal is 7/64 inch thick. A stack of sheet metal is 17 1/2 inches thick. How many sheets are in the stack? __________

28. What is the height of the finished ceiling in this room? __________

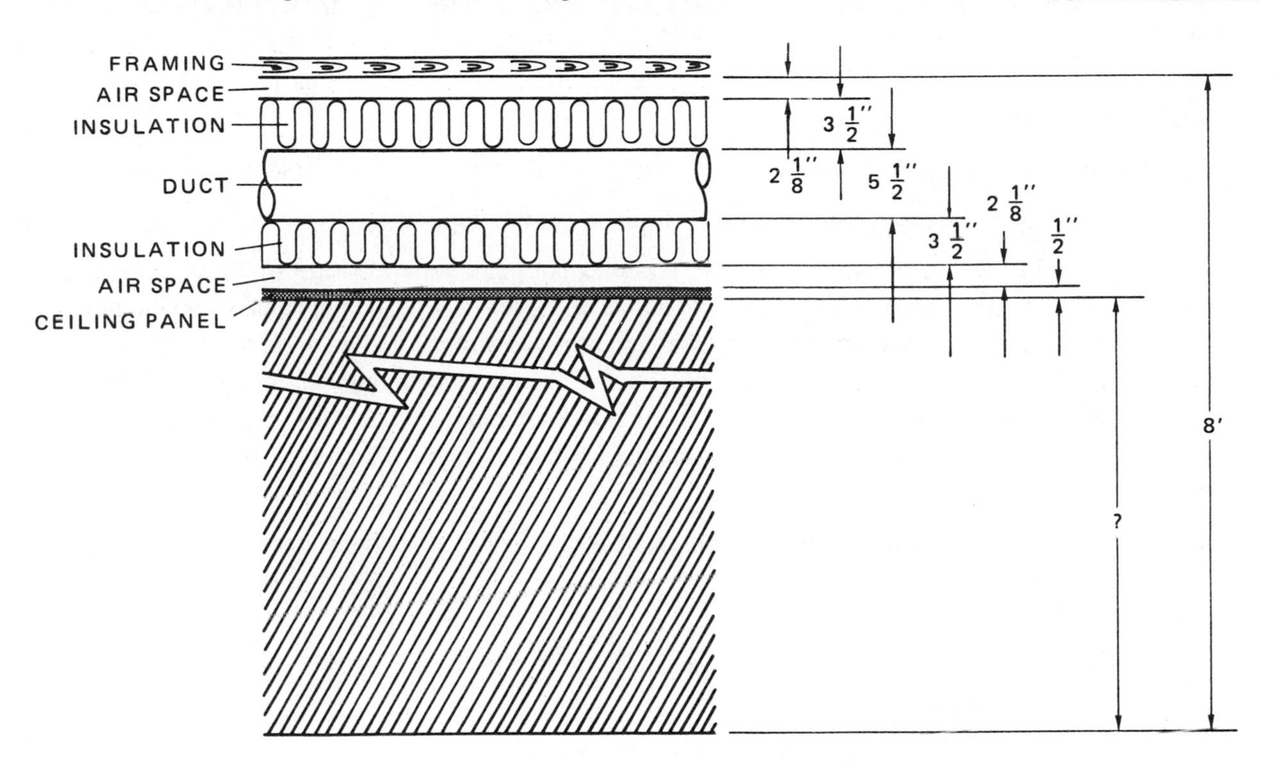

29. A piece of duct measures 24 inches. When 2 pieces are fitted together, 1 1/2 inches are allowed on each duct for fitting. On a job, 27 1/2 pieces of ducting are used. Find, in inches, the length of the finished duct. __________

Decimal Fractions

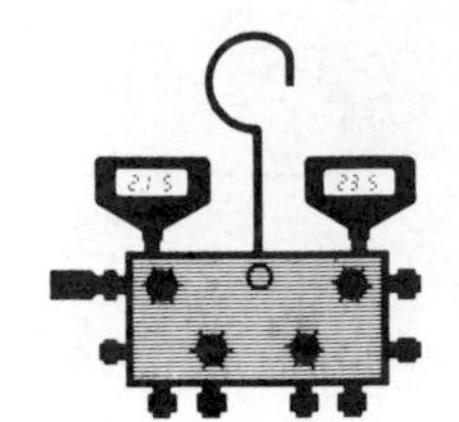

Unit 11 ADDITION OF DECIMAL FRACTIONS

BASIC PRINCIPLES

- Review and apply the principles of addition of decimal fractions to the problems in this unit.
- Study addition of denominate numbers in Section I of the appendix.

A decimal is number with a decimal point in it. The decimal point separates the whole number of the number from the fraction part. The whole number part is to the left of the decimal point, while the fraction part is to the right. Each number position of the fraction part has a name just as each position of a whole number has a name. So the number 123.456 780 9 has 123 as the whole number part and 456 780 9 as the fraction part. Four is in the tenths position, 5 in the hundredths position, 6 in the thousandths position, 7 in the ten thousandths position, 8 in the hundred thousandths position, 0 in the millionths position, and 9 in the ten millionths position.

Adding decimal numbers is just like adding whole numbers, with one additional concern. When writing the numbers, line up the decimal points.

```
233.45
 18.9
506.807
-------
```

Notice that no 0's are written at the end of fraction parts of the numbers. Add the columns just as was done with whole numbers. Carry numbers just as was done with whole numbers. Place the decimal point directly under where it is in the column.

```
 12
233.45
 18.9
506.807
-------
759.157
```

Note: When adding decimal fractions, always be sure that the decimal points are lined up one under the other.

PRACTICAL PROBLEMS

Add.

1. 214.71
172.55
+ 187.37

2. 0.913
8.047
+ 76.465

3. 54.
330.713
+ 8.92

4. 1 731.862 meters
3.14 meters
+ 92.67 meters

5. 0.812 cu in
960.245 cu in
37.043 cu in
+ 251.3 cu in

6. 121.44 square meters
40.08 square meters
5.91 square meters
+ 636.36 square meters

7. 58.81 + 49.97 + 36.54 ________

8. 32.987 + 1 993.25 + 520.09 ________

9. 0.262 3 + 1 750.97 + 6.844 ________

10. 29.97 liters + 0.348 4 liter + 1.822 liters ________

11. 225 lb + 598.17 lb + 1.982 lb ________

12. 409.65 gal + 0.644 gal + 79.7 gal + 69.286 gal ________

13. In a normal refrigerating cycle the vapor pressure for refrigerant R-22 entering the compressor is 68 pounds per square inch gauge reading (psig). The condensing pressure is 131.5 psig higher than the vapor pressure. What is the condensing pressure for refrigerant R-22? ________

14. Absolute pressure = Gauge pressure + Atmospheric pressure
A pressure gauge in a refrigerator system is reading 119.6 pounds per square inch gauge pressure when the atmospheric pressure is 14.69 pounds per square inch. What is the absolute pressure of the refrigerant in that system? ________

15. The gasket on this refrigerator is worn and no longer fills the gap. The gasket is to be replaced. Find the thickness of the new gasket. ________

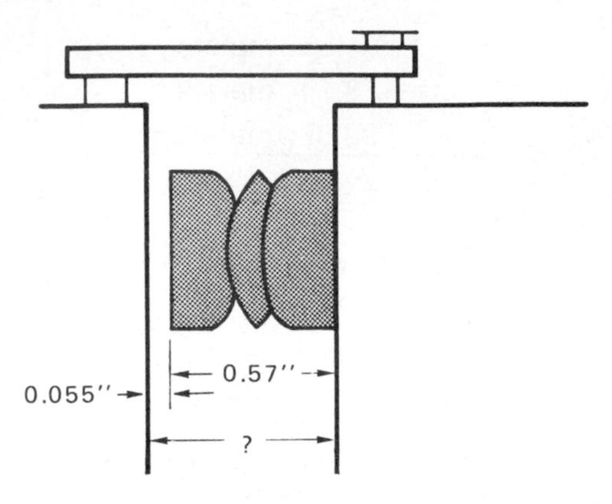

16. A section of a hot water heating system must be suspended. The section is made from one-inch copper tubing weighing 8.515 pounds. The water inside the tubing weighs 4.655 8 pounds. What is the total weight that must be supported in this section? ________

17. In a refrigeration cycle, the refrigerant gains heat in the evaporator and in the suction line. In a certain refrigeration system, R-22 gains 89.8 Btu/lb in the evaporator. It then gains 2.5 Btu/lb in the suction line. Find, in British thermal units per pound, the total heat gained by the R-22. ________

18. In a parallel electrical circuit, the total current is the sum of the individual currents. An air conditioner has a blower and a compressor that are wired in a parallel circuit. The blower draws 1.042 amperes and the compressor draws 11.95 amperes of current. When both are operating, what is the total current drawn? ________

19. What is the outside diameter of this tube? ________

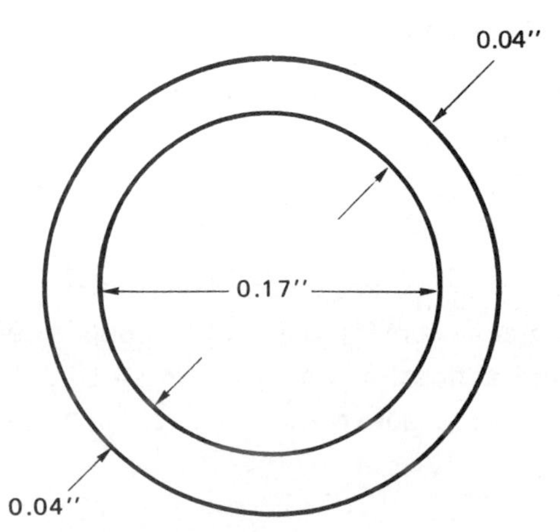

20. A repairer uses these items in fixing a hot water heating system.

ITEM	COST
acid core solder	$ 2.49
solder paste	$ 0.59
stop valve	$ 2.85
90° elbow	$ 0.43
1/2-inch copper tubing	$ 5.34

Find the total cost of the items used. ______

Note: Use this information and diagram for problems 21 and 22.
The *R value* of a substance gives the resistance of a substance to heat flow. The larger the R value, the more resistant the substance is to heat flow and heat loss.

R VALUES FOR CERTAIN MATERIALS

Material	R value
Air space (3 1/2")	0.97
Brick veneer	0.56
Fiberglass insulation (3 1/2")	11.00
Wallboard (interior)	0.79
Wood siding and building paper	0.86

The total resistance is the sum of each substance's resistance.

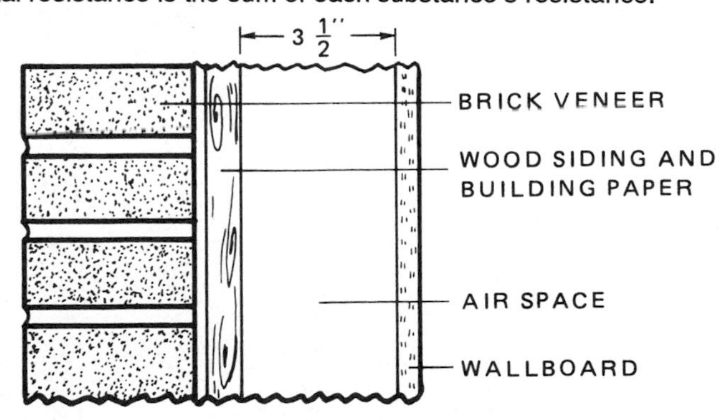

21. What is the total R value for this wall? ______

22. Fiberglass insulation, 3 1/2" thick, is placed in the air space of the wall. Find the R value. ______

23. A partially filled tank of R-22 weighs 47.3 pounds. Another 15.5 pounds of R-22 are put into the tank. How much does the tank now weigh? ______

24. At 86 degrees Fahrenheit (°F) and 172.87 pounds per square inch absolute, R-22 boils (it changes from a liquid to a gas). One pound of liquid R-22 occupies 0.013 65 cubic feet of volume. When it changes to a gas, it requires 0.306 35 cubic feet more space. What volume will one pound of R-22 occupy as a gas at 86°F and 172.87 pounds per square inch? ______

25. Upon being started, a motor draws additional current until the motor is running. A large fan motor draws 3.45 amps of current when running. When starting the fan motor draws an additional 9.27 amps of starting current. What is the total current drawn by the fan motor while starting? ______

Unit 12 SUBTRACTION OF DECIMAL FRACTIONS

BASIC PRINCIPLES

- Review and apply the principles of subtraction of decimal fractions to the problems in this unit.
- Study subtraction of denominate numbers in section 1 of the appendix.

To subtract decimals once again line up the decimal point. Then subtract just as was done with whole numbers. You may have to borrow as was done with whole numbers.

$$\begin{array}{r} 25.3 \\ -\ 8.46 \\ \hline \end{array}$$

Treat the blank after 3 as a 0. Borrow from the 3 and make the 0 a 10. In this problem you will also have to borrow from the 5 to make a 12 above the 4 and from the 2 to make a 14 above the 8.

$$\begin{array}{r} 25.3 \\ -\ 8.46 \\ \hline 16.84 \end{array}$$

Just like in addition of decimals, the decimal point goes under the decimal points that were lined up.

PRACTICAL PROBLEMS

Subtract.

1. $\begin{array}{r} 249.83 \\ -\ 126.41 \\ \hline \end{array}$

2. $\begin{array}{r} 0.936\ 6 \\ -\ 0.171\ 8 \\ \hline \end{array}$

3. $\begin{array}{r} 3.754 \\ -\ 2.46 \\ \hline \end{array}$

4. $\begin{array}{r} 325.19\ \text{cm}^2 \\ -\ 44.053\ \text{cm}^2 \\ \hline \end{array}$

5. $\begin{array}{r} 75.013\ \text{sq in} \\ -\ 32.246\ \text{sq in} \\ \hline \end{array}$

6. $\begin{array}{r} 582.476\ \text{L} \\ -\ 488.827\ \text{L} \\ \hline \end{array}$

7. 259.5 – 169.7

8. 534.93 – 401.46 ________

9. 1 447.755 – 829.74 ________

10. 597.6 sq ft – 321.04 sq ft ________

11. 4.518 9 m^2 – 0.261 6 m^2 ________

12. 110 lb – 67.219 lb ________

Note: Use this table for problems 13–15.

PROPERTIES OF CERTAIN REFRIGERANTS

Refrigerant	Temperature (in °F)	Density (in lb/cu ft)	Saturated Vapor Pressure (in lb/sq in)
R-12	5°F 86°F	90.135 80.671	26.483 108.04
R-22	5°F 86°F	83.277 73.278	42.888 172.87

13. At 5°F, how much denser is R-12 than R-22? ________

14. What is the difference in the saturated vapor pressure for refrigerant R-12 at the two given temperatures? ________

15. How much denser is R-12 at 5°F than at 86°F? ________

16. The bearing in an electric fan motor should be replaced if the rotor hits the stator. When assembled, the distance between the rotor and the stator is 0.03 inch. The bearing is now worn 0.001 75 inch. Find the distance remaining between the rotor and the stator. ________

17. In order for a certain refrigerator door to close tightly, the refrigerator must be leveled up, or shimmed, 0.24 inch. If two shims are used, what is the thickness of the second shim? ________

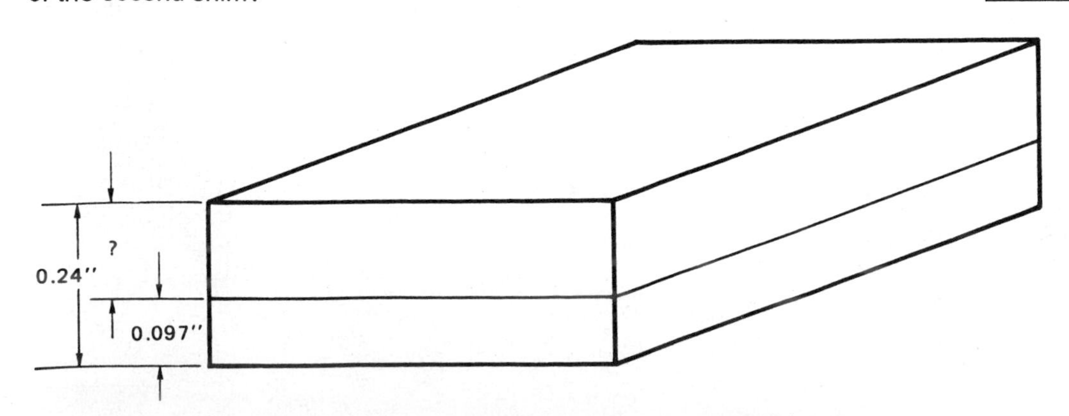

18. At 150°F dry air has a density of 0.065 2 pound per cubic feet. At 70°F it has a density of 0.075 pound per cubic foot. What is the difference in density between air at 70°F and air at 150°F? __________

19. Determine, in pounds per square inch, the pressure drop between the ends of this duct. __________

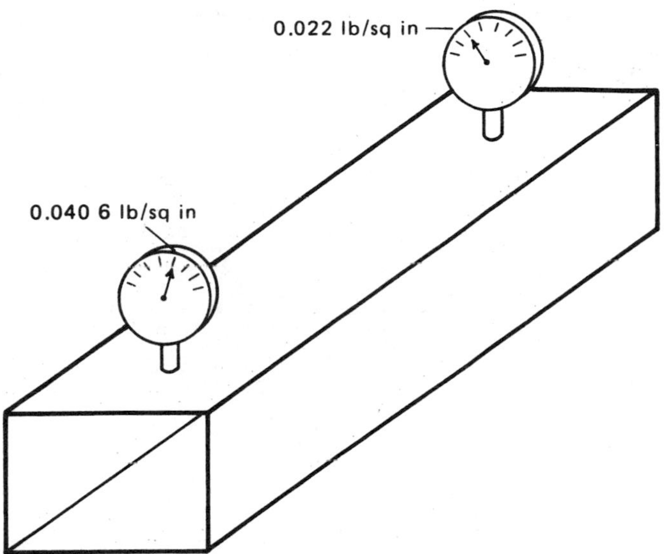

20. At 5°F one pound of refrigerant R-717 vapor has a volume of 8.15 cubic feet. When it is compressed into a liquid, it has a volume of 0.024 cubic foot. Find the change in volume between the vapor and liquid states. __________

21. Find, in inches, the thickness of the insulation in this refrigerator wall. __________

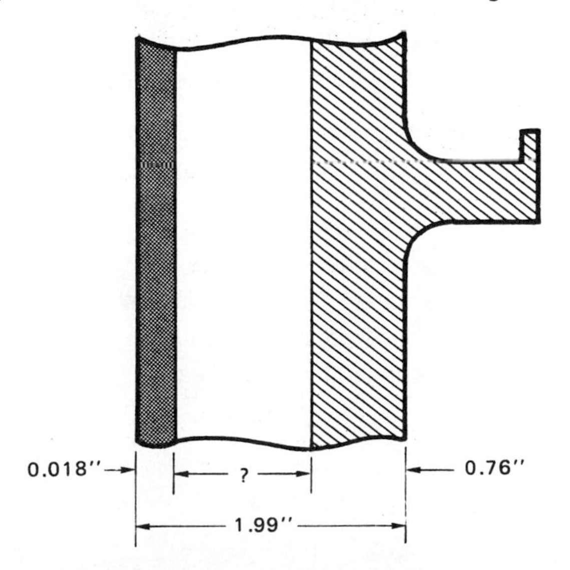

22. The R value of a substance gives the resistance of that substance to heat flow. The larger the R value, the more resistant the substance has to heat flow and heat loss. The total resistance of this wall is 17.85. It is the sum of the R values for each substance. What is the missing R value? __________

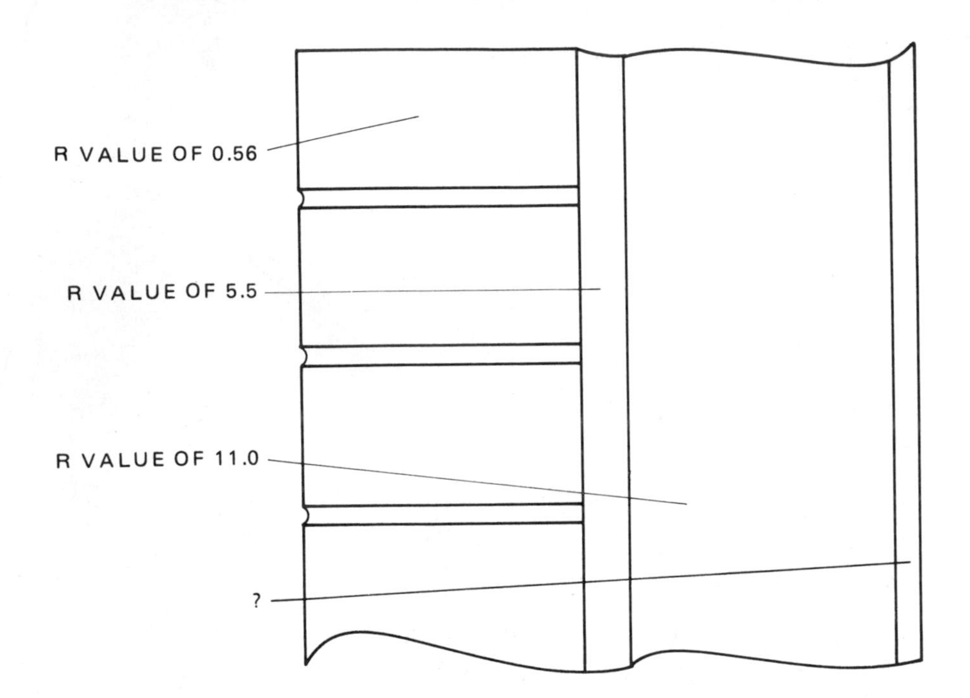

23. A partially filled tank of R-12 weighs 47.3 pounds. When the tank is empty, it weighs 12.6 pounds. What is the weight of the R-12 in the partially filled tank? __________

24. A bill is made out for a completed repair job. The bill is broken down into parts and labor. The total of the bill is $84.57. Parts cost $27.68. What is the cost for labor? __________

25. Absolute pressure = Gauge pressure + Atmospheric pressure
A pressure gauge is being checked for accuracy. The gauge is connected to a tank that has an absolute pressure of 742.11 pounds per square inch (psi). Atmospheric pressure is 14.62 psi. What should the gauge read? __________

Unit 13 MULTIPLICATION OF DECIMAL FRACTIONS

BASIC PRINCIPLES

- Review and apply the principles of multiplication of decimal fractions to the problems in this unit.
- Study multiplication of denominate numbers in section I of the appendix.

When multiplying decimals, do not worry about where the decimal points are. Line up the last numbers. Multiply the numbers just as was done with whole numbers.

```
    51.3
 × 0.469
    4617
   3078
  2052
  240597
```

Now count the total number of places to the right of the decimal point in each of the numbers multiplied together—one in the first number and 3 in the second number. Now count that number of places from the right end of the answer and place the decimal point.

The answer becomes 24.059 7.

A whole number has 0 places to the right of the decimal point. When multiplying a decimal number times a whole number, the answer should have as many decimal places in it as the decimal number in the problem.

Note: When multiplying decimal fractions, the decimals do not have to be lined up. The number of places to the right of the decimal point in the product is equal to the sum of the number of places to the right of the decimal point in both the multiplier and the multiplicand.

PRACTICAL PROBLEMS

Multiply.

1. 503.6 × 4.47
2. 3.594 × 0.219

3. 76.196
× 0.072

4. 48.13 in
× 30.07

5. 0.555 3 lb
× 13.63

6. 31.974 min
× 3.65

7. 8.683 × 0.403 9

8. 490.21 × 3.72

9. 65.2 × 0.000 17

10. 2.398 × 5.3 ft

11. 397 cm^3 × 0.103 1

12. 41.59 L × 34.23

13. In all liquid refrigerant lines, the pressure at the bottom of a vertical rise is greater than the pressure at the top of the rise. For R-12, the pressure is about 0.054 lb/sq in less for every vertical foot of pipe that is used. The vertical rise between this condenser and evaporator is 7.5 feet. How much less is the pressure at the top of the rise than at the bottom?

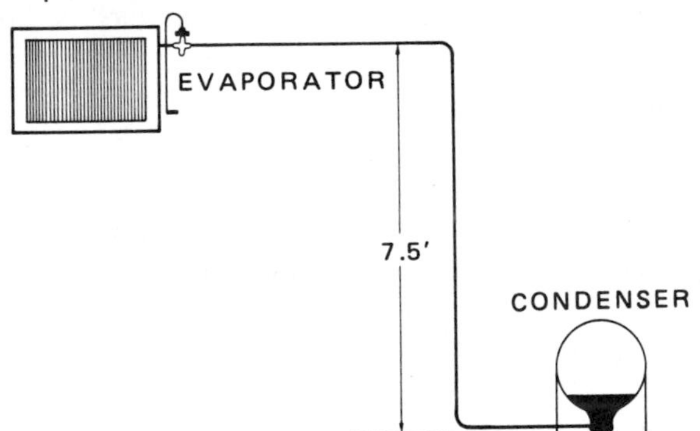

14. The rating stamped on an oil burner nozzle tells the number of gallons of oil sprayed each hour. One day the nozzle sprayed oil for 8.25 hours. How many gallons are sprayed during this day?

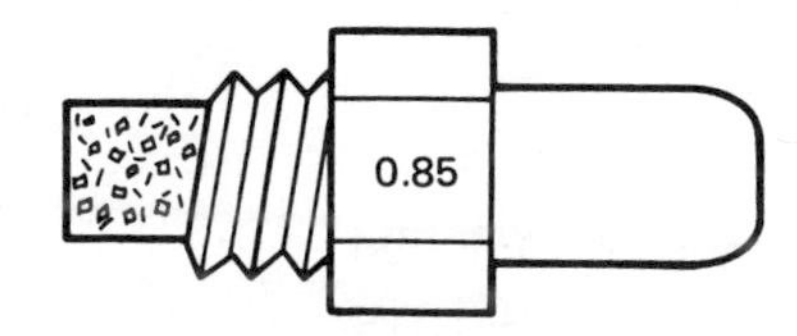

15. A 1/2-inch diameter fuel line runs from an oil tank to the burner. The line is made of 1/2-inch copper tubing weighing 0.344 pound per foot. What is the total weight of the tubing? __________

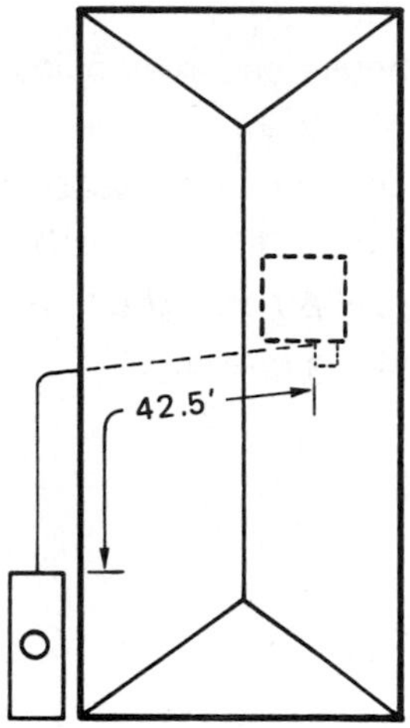

16. At a certain house, 147.3 gallons of #2 fuel oil are unloaded. The fuel oil weighs 7.1 pounds per gallon. Find, in pounds, the total weight of the fuel oil unloaded. __________

17. A mullion heater prevents condensation on the refrigerator cabinet between the two doors of the cabinet. One mullion heater has a value of 12.5 watts. One watt produces 3.415 Btu (British thermal units) of heat. How much heat does the mullion heater produce? __________

18. A repairer's assistant earns $6.20 per hour. How much is the assistant paid for a job lasting 4.5 hours? __________

19. When one pound of refrigerant R-22 vaporizes, 93.21 Btu of heat are removed from the surroundings. How many British thermal units of heat are removed when 9.4 pounds of R-22 vaporize?

20. The distance from the shop to the work site is 7.6 miles. An installer must make 4 round trips while installing a new heating system. How much mileage should be recorded for this job? __________

21. The manufacturer's manual for a fan motor states that the motor draws a starting current that is 6.2 times larger than its running current. If its running current is 2.147 amperes, find the expected current reading on an ammeter when starting the fan. __________

22. If a room that was used as part of a house becomes office space and is air conditioned rather than just heated as the house was, the number of air changes per hour is increased. For the room in question, the number of air changes per hour became 2.3 times larger. If the old number of changes was 3.4 changes per hour and the room was 952.7 cubic feet, what flow should the new ventilating system be able to handle in one hour? __________

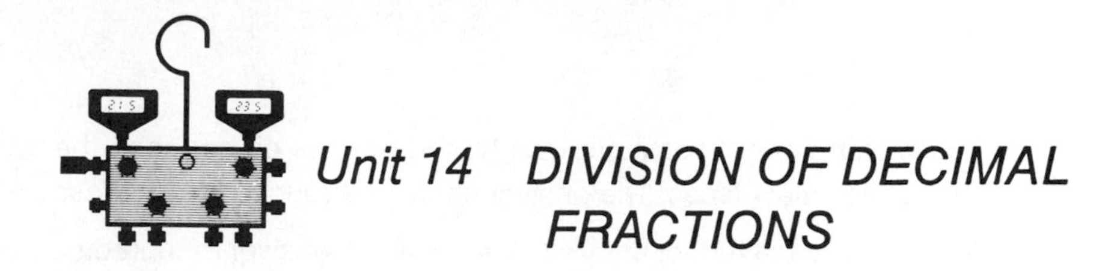

Unit 14 DIVISION OF DECIMAL FRACTIONS

BASIC PRINCIPLES

- Review and apply the principles of division of decimal fractions to the problems in this unit.
- Study division of denominate numbers in section I of the appendix.

There are a few differences between dividing whole numbers and dividing decimal numbers. The problems are still written in the same way.

$$1.65\overline{)5.742}$$

But, before dividing, move the decimal point in the divisor so that it is now a whole number. In this case that would be 2 places to the right. Then do the same thing in the dividend, and place the decimal point directly above that point in the quotient.

$$1\,65.\overline{)5\,74.2}$$

Now ignore the decimal point and divide as if the numbers were whole numbers; be careful to place the number in the quotient in the correct place.

$$\begin{array}{r} 3.4 \\ 1\,65.\overline{)5\,74.2} \\ \underline{4\,95} \\ +\,79\,2 \\ \underline{66\,0} \\ 13\,2 \end{array}$$

Now we can add 0's onto the end of the dividend to carry the division out further.

$$\begin{array}{r} 3.48 \\ 1\,65.\overline{)5\,74.20} \\ \underline{4\,95} \\ +\,79\,2 \\ \underline{66\,0} \\ 13\,20 \\ \underline{13\,20} \end{array}$$

There are times when the division will not come out evenly, or when we do not need the answer to be more than a certain number of decimal places. The answer should be rounded off in these cases.

To round off, carry the division one place further than is asked for in the answer (hundredths if tenths were called for, thousandths if hundredths were called for, etc.). Look at the last number. If it is less than 5, drop the last number and you have your answer. If the number is 5 or larger, drop the last number, but raise the next number by 1 to give you the answer. For example, 3.24 rounded off to the nearer tenth would be 3.2 (since 4 is less than 5); 151.348 rounded off to the nearer hundredth would be 151.35 (8 is larger than 5, so drop the 8 and make the 4 a 5).

Note: When dividing decimals long hand, move the decimal point to the right to make the divisor a whole number. Move the decimal point the same number of places to the right in the dividend and place it above that point in the quotient.

PRACTICAL PROBLEMS

Divide.

1. 17)59.84
2. 5.3)343.175
3. 0.375)0.156 75
4. 8.09)760.783 6 cu in
5.  0.002 61)0.019 632 42 gal
6. 6.17)0.466 708 m
7. 2 073 ÷ 3.9 (Round the answer to the nearer tenth.)
8. 31.71 ÷ 0.075 5
9. 2 844.686 ÷ 923
10. 9.465 712 cu yd ÷ 4.714
11. 937.135 in ÷ 28.1 in 
12.  0.787 826 L ÷ 0.001 3 
13. A repairer is troubleshooting a problem in an electrical circuit. Seven identical air conditioners are running on one circuit. With all of them running, 12.845 amperes of current flow through the circuit. What should each air conditioner have as current running through its unit? 
14. A 50-foot coil of 3/8-inch diameter copper tubing weighs 9.9 pounds. What is the weight of 1 foot of the tubing? 

15. Washers used for 3/16-inch bolts weigh 0.21 pound per box. The box contains 75 washers. Find, to the nearer thousandth pound, the weight of each washer. ______

16. A stack of duct metal is 4 inches high. If there are 128 sheets in the stack, what is the thickness of each sheet? ______

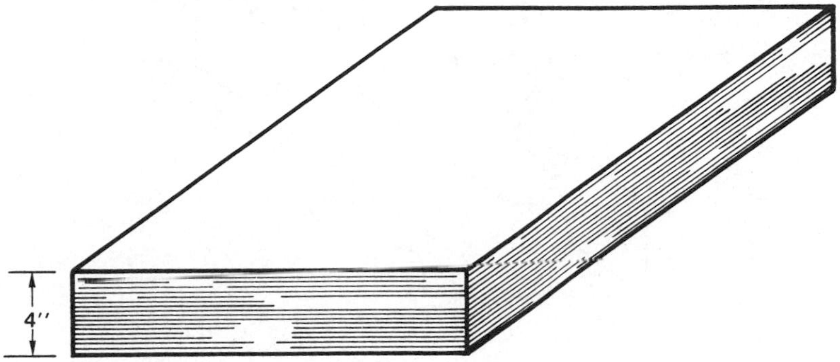

4"

17. An oil burner ran a total of 4.5 hours in one day and used 7.425 gallons of fuel. How many gallons would be used if it ran only one hour? ______

18. A repairer is paid $57.51 for a job. The job takes 6.75 hours. What is the repairer's hourly rate? ______

19. The degree to which heat flows through a substance is called *thermal conductivity*. Thermal conductivity of a substance is given in U values. A U value is based on British thermal units, area in square feet, time in hours, and temperature change in degrees Fahrenheit (Btu/sq ft · h · °F). The U factor is 1 divided by the thermal resistance, or R value. The R value for a house is 3.51. What is the U value? Round the answer to the nearer ten-thousandth. ______

20. The *density* of a substance is the weight of the substance divided by its volume. At 5°F, the weight of 3.5 cubic feet of liquid R-12 is 315.49 pounds. What is the density in pounds per cubic foot? ______

21. When a refrigerant vaporizes, it takes heat from its surroundings. The *latent heat of vaporization* is the amount of heat needed for the refrigerant to vaporize. When 38.5 pounds of R-502 vaporize at 5°F, the latent heat of vaporization is 2 651.11 Btu. What is the latent heat of vaporization for 1 pound of R-502 at 5°F? ______

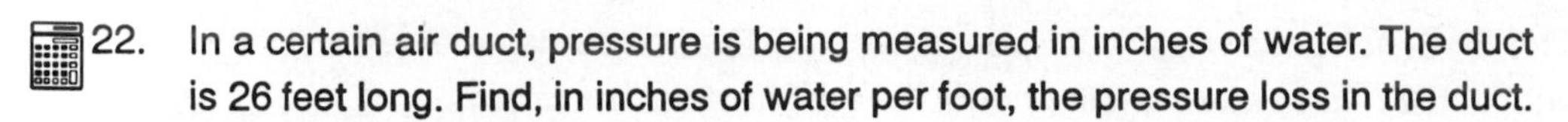

22. In a certain air duct, pressure is being measured in inches of water. The duct is 26 feet long. Find, in inches of water per foot, the pressure loss in the duct. ________

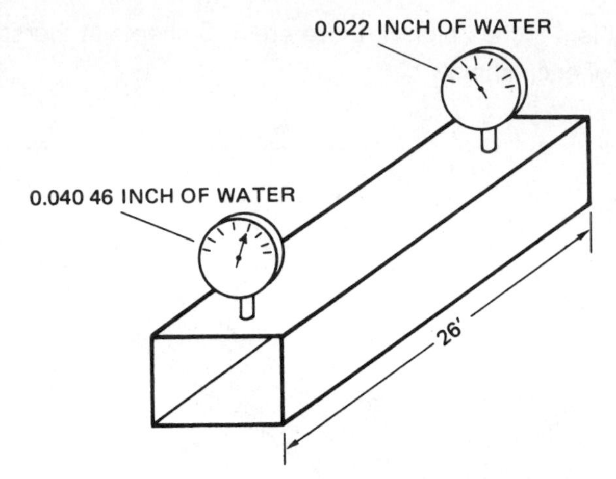

23. Find the wall thickness of this rigid polyvinylchloride (PVC) tube. ________

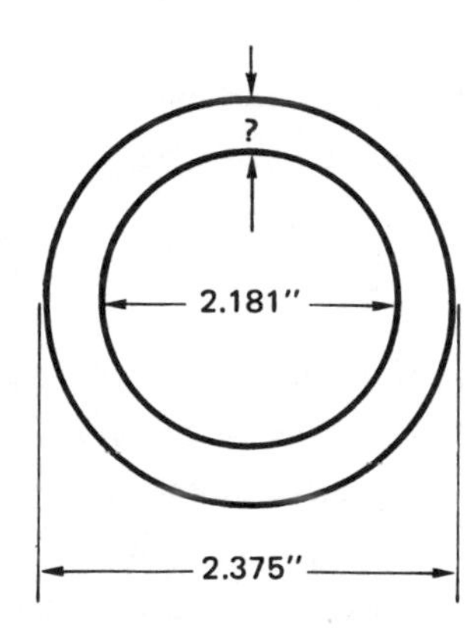

24. It took the Keep Kool Company 32.75 hours to install identical air conditioning ducts in 6 new condominium units. What is the time it took to install ducts in one unit? ________

25. A circulating fan can supply 9 525.6 cubic feet of air in one hour. What is the largest volume room that will have its air exchanged 4.5 times each hour? ________

Unit 15 DECIMAL AND COMMON FRACTION EQUIVALENTS

BASIC PRINCIPLES

- Review and apply the principles of decimal and common fraction equivalents to the problems in this unit.
- Study denominate numbers in section I of the appendix.

Decimal and common fractions are both types of fractions. Therefore there should be some way of getting from one type of fraction to the other.

To convert from a decimal to a common fraction, take the number to the right of the decimal point and make that the numerator of the common fraction. The denominator is the value of the place of the last number of the decimal.

Thus, 0.5 is five tenths. The fractional equivalent is $\frac{5}{10}$. This fraction, like all other fractions, could be reduced. This one can be reduced to $\frac{1}{2}$.

Another example is 0.431 or 431 thousandths, which has the common fractional equivalent of $\frac{431}{1000}$.

To convert from a common fraction to a decimal, divide the denominator into the numerator and carry the answer out as a decimal.

Therefore, $\frac{1}{4}$ can be converted to a decimal by dividing 1 by 4.

$$\begin{array}{r} 0.25 \\ 4\overline{)1.00} \\ \underline{8} \\ 20 \\ \underline{20} \end{array}$$

A problem that can occur when doing this process is that many common fractions do not have even decimal equivalents. So when the decimal starts to repeat itself, you may want to carry it out just a couple of decimal places unless otherwise instructed.

Therefore, the decimal equivalent of $\frac{1}{3}$ is 0.33 and the equivalent of $\frac{1}{9}$ is 0.11.

Note: To change any fraction into a decimal, divide the denominator into the numerator. Many fractions do not have exact decimal equivalents. They have to be rounded off.

PRACTICAL PROBLEMS

Express each common fraction as a decimal fraction.

1. 3/16 __________
2. 5/64 __________
3. 7/8 __________
4. 8/25 __________
5. 7/40 __________

Express each common fraction as a decimal fraction. Round each answer to four decimal places.

6. 1/3 __________
7. 4/15 __________
8. 3/13 __________
9. 1/6 __________
10. 5/11 __________

Do the following mathematical operations by first converting the fraction to a decimal number and then performing the indicated operation.

11. $2\frac{5}{8} + .0312$
12. $4\frac{1}{4}$ inches – 1.264 inches
13. $117.264 \times 3\frac{3}{4}$

Round off the following problems to 3 decimal places.

14. 21.42 ounces + $2\frac{2}{9}$ ounces __________
15. $64\frac{7}{8} \times 7.21$ __________
16. $3\frac{1}{5} + 7\frac{2}{7} + 43.97$ __________

17. Find, in inches, the inside diameter of this water pipe for an automatic ice maker. Express the answer as a decimal fraction. __________

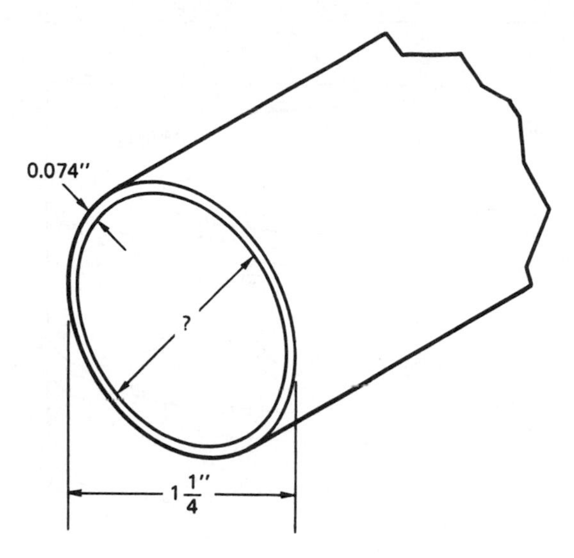

18. The gap between the electrodes of an oil burner gun is set at 1/8 inch. Use has caused each electrode to wear 0.015 5 inch. What is the new gap distance expressed as a decimal fraction? __________

19. The slight sidewards motion of a shaft is called *end play*. The end play in the shaft of a rotor for an electric motor should not be more than 1/32 inch. One motor has an end play of 0.030 5 inch. Is this end play more than 1/32 inch? __________

20. A compressor has a volume of 1 5/8 cubic inches. On each stroke the compressor pumps 0.63 times its volume. What volume does the compressor pump on each stroke? Express the answer as a decimal fraction. __________

21. What is the outside diameter of this piston ring, expressed as a decimal fraction? ________

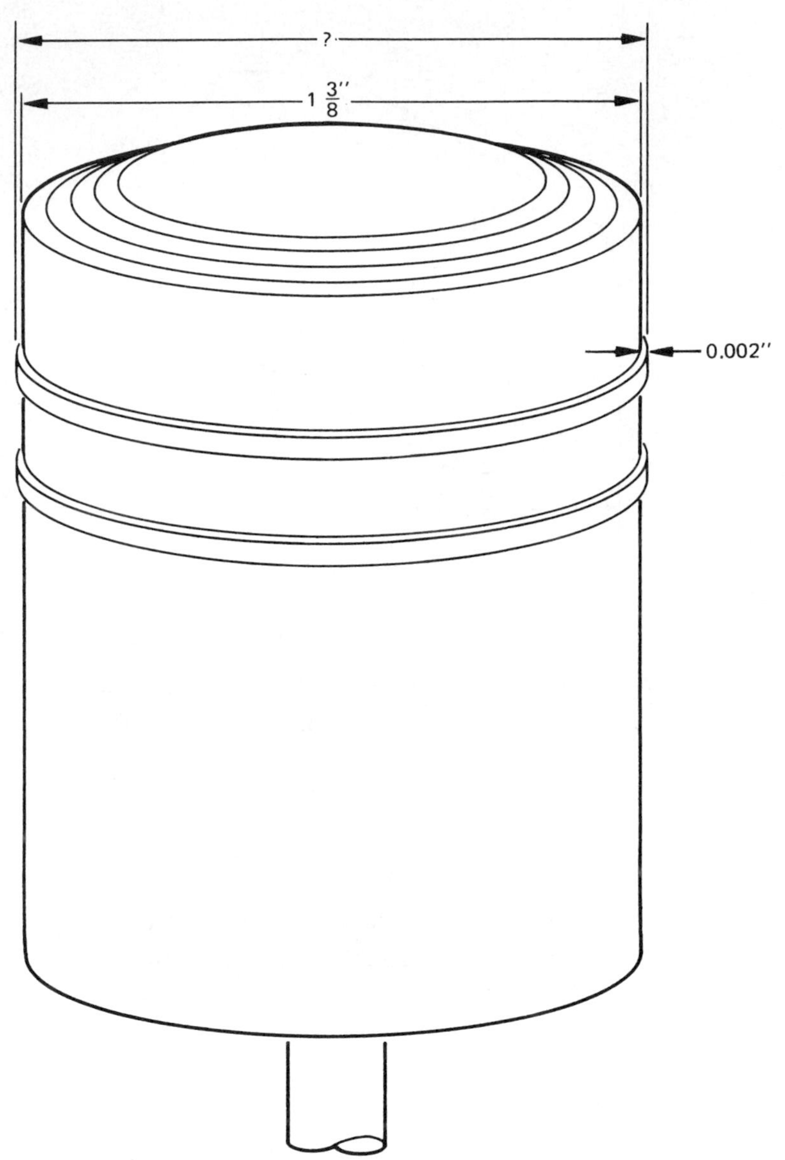

22. A major repair job on a refrigeration compressor took 16.5 hours to complete. Joe worked 1/2 of that time, Bill worked 1/3, while Al worked 1/6 of the total. What time, expressed as a decimal, should be put on each worker's payroll sheet? Round off each answer to the nearer tenth. ________

23. When an object goes on sale, the savings can be found by multiplying the portion taken off by the price of the object.
    A heating supply dealer has taken 1/5 off the price of an acetylene torch. If the price was $233.57, what is the savings for buying the torch today? __________

24. An electrical store gives 1/8 off the price of a crimper for making electrical connections when 6 or more are purchased. What would be the savings on a $29.95 crimper if at least 6 are bought? Round off to the nearer cent. __________

25. A humidifier is designed to put 8.3 gal of water into air flowing through an air duct every 24 hours when running continuously. How much water is put into the air when the system runs only 2/5 of the time? __________

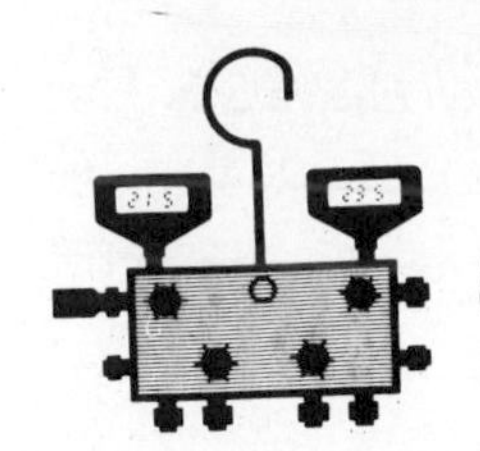

Unit 16 COMBINED OPERATIONS WITH DECIMAL FRACTIONS

BASIC PRINCIPLES

- Review and apply the principles of addition, subtraction, multiplication, and division of decimal fractions to these problems.

PRACTICAL PROBLEMS

1. $\begin{array}{r} 17.315 \\ 2.804\,7 \\ +\ 0.766\,4 \\ \hline \end{array}$

2. $\begin{array}{r} 0.505 \\ 23.0 \\ +\ 6.821\,7 \\ \hline \end{array}$

3. $\begin{array}{r} 15.768 \\ -\ 3.503 \\ \hline \end{array}$

4. $\begin{array}{r} 7.244 \\ -\ 3.503 \\ \hline \end{array}$

5. $\begin{array}{r} 16.23 \\ \times\ 0.683 \\ \hline \end{array}$

6. $\begin{array}{r} 8.905 \\ \times\ 0.004\,9 \\ \hline \end{array}$

7. $8.41 \overline{)\ 38.013\,2}$

8. $0.002\,89 \overline{)\ 0.000\,109\,82}$

9. 4.437 cm + 0.091 8 cm + 85.68 cm ______
10. 0.005 7 gal + 7.899 4 gal + 3.013 gal + 11.652 gal ______

11. 13.1 ft – 5.207 ft ______
12. 6.349 22 sq yd – 3.707 3 sq yd ______
13. 351.6 m × 0.054 5 ______
14. 0.057 1 × 6.206 L ______
15. 1.338 3 cu in ÷ 0.527 cu in ______
16. 39 lb ÷ 0.704 (Round the answer to the nearest tenth.)

Express each common fraction as a decimal fraction. Round to four decimal places when needed.

17. 3/8 ________

18. 2/5 ________

19. 3/20 ________

20. 5/7 ________

21. What is the inside diameter of this pipe? ________

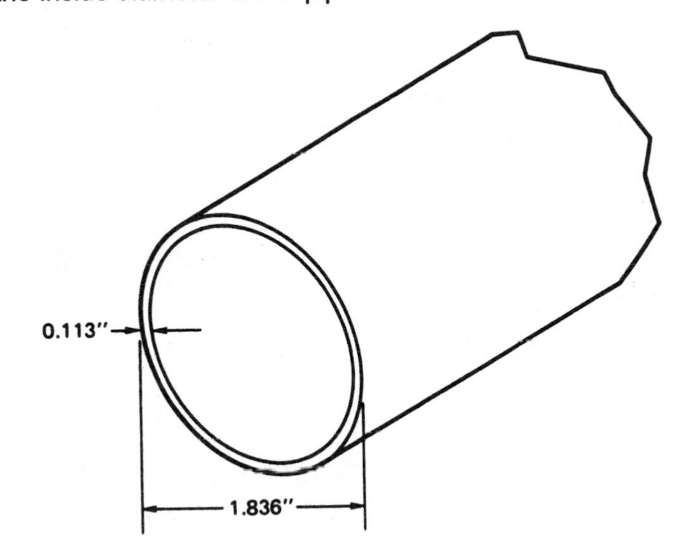

22. At 5°F the latent heat of vaporization for R-12 is 68.204 Btu/lb. This means that when 1 pound of R-12 vaporizes at 5°F, 68.204 Btu of heat are removed from the surroundings. How many pounds of R-12 must vaporize at 5°F in order to remove 13 961.36 Btu of heat from the surroundings? ________

23. At 86°F the density of liquid refrigerant R-22 is 73.278 lb/cu ft. This means that at 86°F 1 cubic foot of liquid R-22 weighs 73.278 pounds. At 86°F, how much would 1.25 cubic feet of liquid R-22 weigh? ________

24. At 10°F the heat content of one pound of liquid refrigerant R-500 is 12.23 Btu. The latent heat of vaporization for one pound of R-500 at 10°F is 81.8 Btu. What is the heat content of one pound of R-500 vapor at 10°F? ________

25. A coil of copper refrigerator tubing weighs 6.275 pounds. The coil is 25 feet long. What is the weight of 1 foot of tubing? ________

26. This tube is to be used as a conduit for wires to a condenser unit. The conduit must pass through a wall. A 25/32-inch drill is used to make the hole. Find the clearance around the tube. Express the answer as a decimal fraction. __________

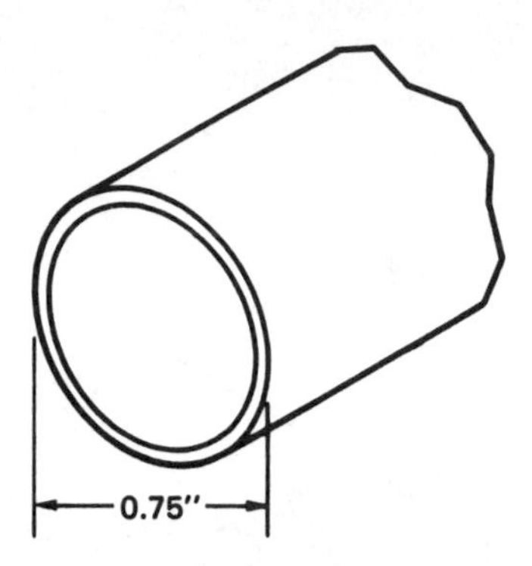

27. What is the total thickness of this refrigerator wall? __________

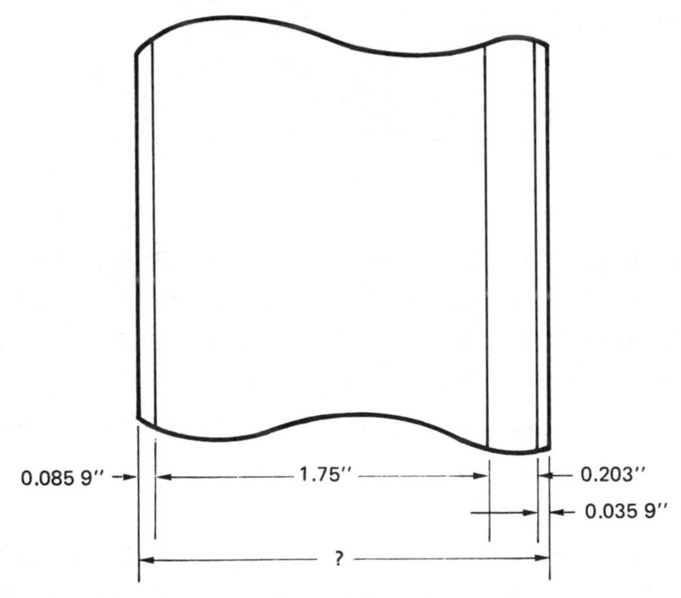

28. At 5°F the latent heat of vaporization for R-12 is 68.204 Btu/lb and 565 Btu/lb for R-717. Find, in British thermal units per pound, the difference in these two values. __________

29. An air duct is made from *26* gauge metal which is 0.017 9 inch thick. The inside width of the duct is 8 3/16 inches. Find the outside width of the duct. Express the answer as a decimal fraction. __________

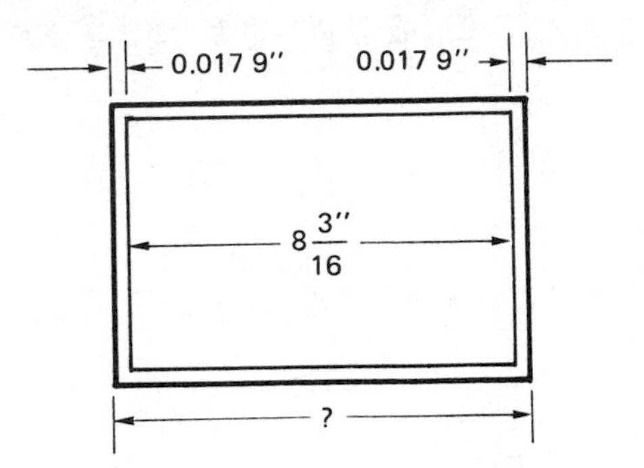

30. At 5°F the latent heat of vaporization for refrigerant R-22 is 93.21 Btu/lb. How many British thermal units of heat are needed when 13.4 pounds of R-22 are vaporized? __________

Percent, Percentage, and Discount

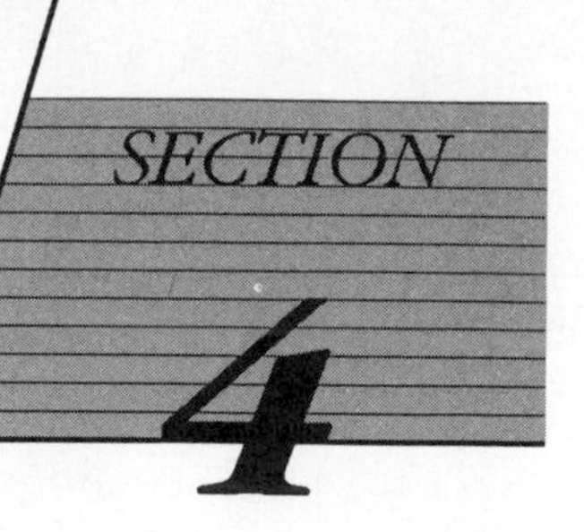

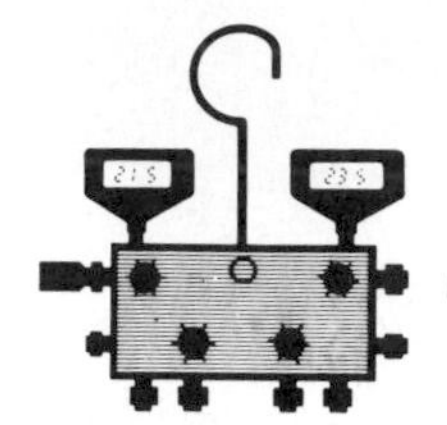

Unit 17 PERCENT AND PERCENTAGE

BASIC PRINCIPLES

- Review and apply the principles of percent and percentage to the problems in this unit.
- Review denominate numbers in section I of the appendix.

Percent is a method of writing decimals as whole numbers. The percent sign (%) takes the place of two decimal places. A percent is usually given with the percent sign. When doing a mathematical operation with the percent, it should be as a decimal number. To change the number from a percent to a decimal, drop the percent sign and move the decimal point two places to the left; 35% becomes 0.35, while 50% becomes 0.5 (you drop 0's at the end of decimals). The decimal point is always moved two places to the left—so, 2% becomes 0.02 (a 0 had to be added in front of the 2 since there were not two places to move the decimal point).

When writing a decimal as a number with a percent sign, just do the opposite to what was done above. Move the decimal point two places to the right and add the percent sign. Convert a common fraction to its decimal equivalent before writing it with a percent.

When trying to find the percent, form a fraction of $\frac{\text{part}}{\text{whole amount}}$ and find the decimal equivalent of this. Then convert it to a number with the percent sign.

Almost all percentage problems can be written in the form:

Some % **of** a number **is** something else

The statement can always be rewritten as a mathematical problem by putting a × sign in place of the word **of** and an = sign in place of the word **is**. The problem can be presented three different ways:

1. Some % of a number is What.
2. Something is some % of What.
3. Something is What % of a number.

These are solved as follows:

1. Convert the % to a decimal and then multiply that times the number to get the answer.
2. Convert the % to a decimal and then divide it into the number to get the answer.
3. Divide the "something" number by the other number and convert the decimal number to a percent number.

Use the following guidelines when figuring percent problems:

- When converting from a percent to a decimal, drop the percent sign and move the decimal point 2 places to the left.
- A large number of percent problems can be written as

  Some % of a number is another number

This form can be quickly written as a mathematical equivalent by:

1. Replace the % by its equivalent decimal.
2. Putting a × sign in place of the word "of."
3. Putting an = sign in place of the word "is."

PRACTICAL PROBLEMS

Express each percent as a decimal number.

1. 2% __________
2. 7.2% __________
3. 12% __________
4. 35 1/2% __________
5. 83.7% __________
6. 125% __________

7. What is 5% of 83? __________
8. Find 27% of 115. __________
9. What percent of 175 is 250? Round the answer to the nearer whole percent. __________
10. The number 45 is what percent of 70? Round the answer to the nearer whole percent. __________
11. The number 24 is 15% of what number? __________
12. The number 125 is 70% of what number? Round the answer to the nearer hundredth. __________

13. An air conditioning installer works part time and has a taxable income of \$6 520. The state income tax is 7% of the taxable income. How much money does the installer pay in state taxes? ______

14. In one day, the blower fan of a forced air heating system runs for 9 hours. What percent of the day does the fan run? Round the answer to the nearer whole percent. ______

15. During a 40-hour workweek a repairer spends 15% of the time driving to and from various jobs. How many hours are spent driving? ______

16. In order to provide power to a window-mounted air conditioning unit, 60 feet of electrical cable are used. This is 20% of the entire reel of cable. Find the number of feet of cable that the reel holds. ______

17. A repair company borrows money to purchase new trucks. The interest paid on the loan is \$1 440. This is 6% of the loan. How much money is borrowed? ______

18. A 250-gallon fuel oil tank has 100 gallons of fuel left in it. What percent of the tank is full? ______

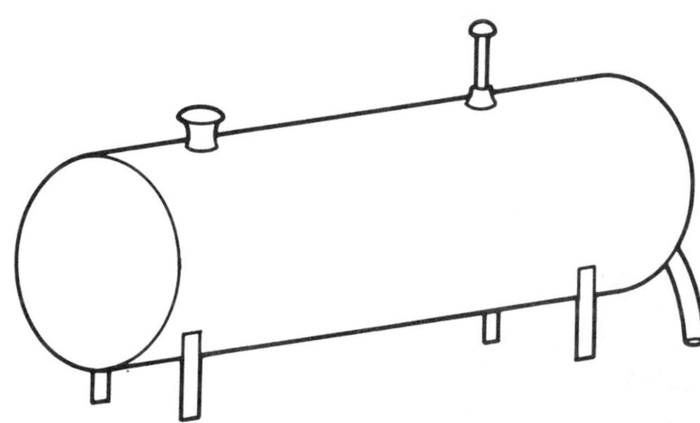

19. In charging a car air conditioner, 24% of a 15-pound cylinder of refrigerant R-12 is used. How many pounds of R-12 are used? ______

20. In making a section of an air duct, 18 square feet of sheet metal are used. This section is cut from a 24-square foot sheet. What percent of the sheet is used? ________

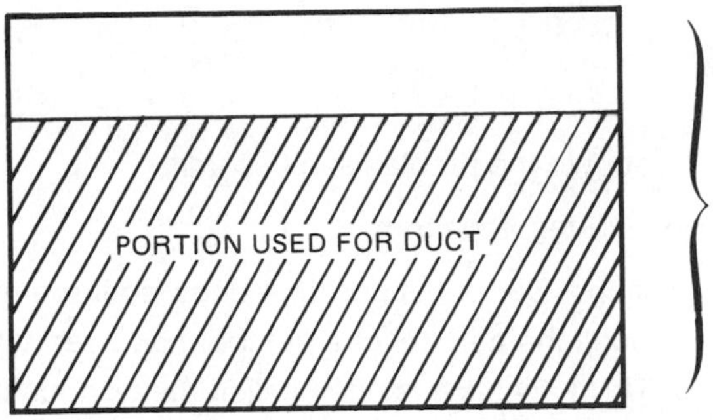

21. A repairer charges $40 for fixing a home heating system. This is 20% of the total bill. Find the amount of the total bill. ________

22. An installer works for 6 hours and completes 48% of a job. How many hours are needed for the complete installation? ________

23. In one hour, the air in a certain room is completely replaced 4 times. What percent of the air is replaced in 5 minutes? Round the answer to the nearer whole percent. ________

24. An air conditioning system is installed in a building under construction. The bill for installation is the cost of the parts plus overhead. The overhead is 75% of the cost of the parts. The parts cost $3 500.
    a. Find the cost of the overhead. a. ________
    b. Find the total amount of the bill. b. ________

25. An air conditioning contractor is installing a heat pump system in a private home. He has agreed to do the job for the cost of parts and labor plus 9%. The heat pump cost $1 195, the electrical supplies $53, the ducting $81, and the other supplies $39. Three men worked 16 hours each to install the system. They were paid $12 per hour.
    a. What is the total cost for parts and labor? a. ________
    b. What is the contractor's fee? b. ________
    c. What is the total bill to the homeowner? c. ________

26. A heating supply store buys 6-inch metal flue pipe at $2.90 per 24-inch section. The pipe is marked up 40%. What is the selling price of one section? ________

27. A heating contractor, in order to cover his expenses, sells parts with a 37% markup. If he bought 1/4-inch copper fuel line tubing for $.38 a foot, what would be its selling price? Round off to the nearer cent. ________

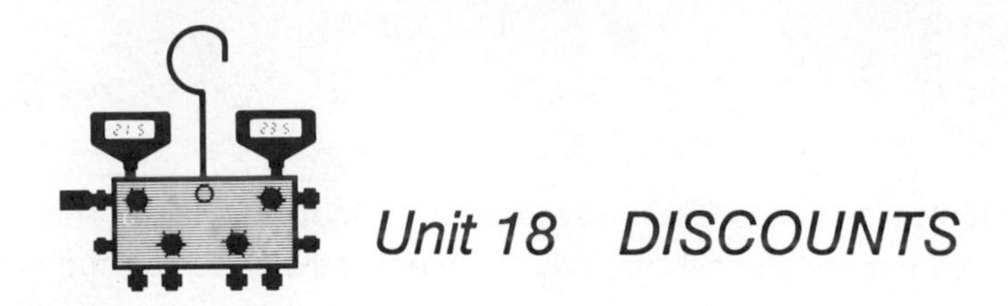

Unit 18 DISCOUNTS

BASIC PRINCIPLES

- Review and apply the principles of discount to the problems in this unit.
- Review denominate numbers in section I of the appendix.

Discounts are percent problems with two exceptions. Discounts deal in money, so there will be decimals in both numbers. The second exception is that once the percent of a number has been taken, that value usually gets subtracted from the original price. This new price is often called the discounted price.

In most discount problems, the answer wanted is the new price. This is found by:

1. Multiplying the old price times the decimal equivalent of the percent. This gives the discount.
2. Subtracting the discount from the old price will give the new (lower) price.

PRACTICAL PROBLEMS

1. A repairer buys replacement parts needed to fix an air conditioning system. The cost of the parts is $53.10. The distributor gives a 12% discount. What is the price the repairer pays? Round the answer to the nearer whole cent. __________
2. A repair company needs parts for an oil burner gun. A distributor will charge $173.40. Another repair company will sell the parts at a 6% discount. If bought from the other repair company, what is the cost of the parts? Round the answer to the nearer whole cent. __________
3. During a sale, insulation for heating ducts is purchased from a distributor. The insulation is valued at $675.00, and there is a 25% discount. Find the sale price of the insulation. __________
4. A new compressor for a refrigerator costs $141.40. An 8% discount is given. Find, to the nearer whole cent, the cost of the compressor. __________
5. A 5 500-Btu/h room air conditioning unit costs $282.00. If three or more units are purchased, a 12% discount is given. What will be the cost of four units? __________
6. A hot water heating system is installed in a house. The system needs 163 feet of 3/4-inch copper tubing. The tubing lists for $0.71 per foot. A 17% discount is given. How much does the tubing cost? Round the answer to the nearer whole cent. __________

7. A residential heat pump system is installed. The bill for the system is $3 950.00. If the bill is paid within 10 days, an 11% discount is given. If the bill is paid after the 10 days and before 30 days, a 5% discount is given.
   a. What is the cost of the system if the bill is paid on the 8th day? a. ______
   b. What is the cost of the system if the bill is paid on the 21st day? b. ______

8. A supply company offers discounts to contractors. The amounts of the discounts are: electrical supplies, 12%; ducts, 9%; plumbing supplies, 11%; hardware, 7%. This is an air conditioning contractor's bill before discounts. ______

THE COMPLETE SUPPLY COMPANY "We have what you need!"		
ITEM	LIST	DISCOUNTED PRICE
300 ft #10 wire	$126.00	
250 ft of 4″ x 8″ ducts	687.50	
25 – 20 ft lengths 3/4″ copper tubing (rigid)	395.00	
6 boxes 1 x 8 sheet metal screws	32.40	
	TOTAL	
Thank you!		

Find, to the near cent, each discounted price.
   a. *#10* wire a. ______
   b. ducts b. ______
   c. copper tubing c. ______
   d. sheet metal screws d. ______
   e. What is the total amount of the bill after the discount? e. ______

9. A department store has a freezer which does not work. The original price of the freezer is $319.50. It is sold at a 45% discount. A refrigeration repairer buys it and spends $32.45 in fixing it. How much does the refrigerator finally cost the repairer? Round the answer to the nearer whole cent.

10. A repair company needs parts for a window air conditioner. The parts list for $273.75. A local supply company gives a repair company a discount of 8%. A distributor in a city 75 miles away gives a discount of 20%. The round trip cost to the city is $30.50.
   a. If the parts are bought from the local company, what is the cost? Round the answer to the nearer whole cent. a. ______
   b. If the parts are bought from the distributor, what is the cost? Round the answer to the nearer whole cent. b. ______
   c. Which place is less expensive? c. ______

11. A shop needs 480 pounds of refrigerant R-12. A supplier charges $0.93 per pound. If the refrigerant is ordered in 60-pound cylinders, a 13% discount is given. If ordered in 15-pound cylinders, a 9% discount is given.
    a. What is the cost of 480 pounds of R-12 if it is ordered in 60-pound cylinders? Round the answer to the nearer whole cent. a. ____________
    b. What is the cost of 480 pounds of R-12 if it is ordered in 15-pound cylinders? Round the answer to the nearer whole cent. b. ____________

Note: Use this information for problems 12–14.
First discount price = List price – (list price × first discount)
Final price = First discount price – (first discount price × second discount)

12. A gas furnace is purchased for a new home. The price is $469.00. The supplier gives discounts of 11% and 3%. Find, to the nearer whole cent, the final cost of the furnace. ____________

13. When purchasing a heat pump unit, an installer is given discounts of 12% and 2%. The unit is priced at $3 400.00. What is the final price of the pump? ____________

14. A shop buys nuts, bolts, washers, and electrical staples from a distributor. The cost of these supplies is $123.72. The distributor gives discounts of 9% and 6%. How much does the shop pay for the supplies? Round the answer to the nearer whole cent. ____________

15. The motor for a commercial condensing unit needs to be replaced. The cost of a new one is $239.00. If the bad motor is turned in, a 33% discount is given. What is the price for the motor if the bad one is turned in? ____________

16. The Toasty Home Heating Company charges $43.50 to tune up an oil burner. During the months of July and August, a 12% discount is given to customers. How much is saved by getting a tune-up in July or August? ____________

17. A heating and air conditioning repairman is buying electrical supplies from the Big Charge Electrical Supply Company. If the bill is paid in cash, a 4% discount is given. How much does the repairer save by paying cash if his bill is $317.25? ____________

18. The owner of the Old Reliable Plumbing and Heating Store has an electric heater he wants to put on sale. Its current price is $79.95. If he gives a 41% discount, he will sell the heater for what he paid for it. What did the owner pay to buy the heater? Round the answer to the nearer whole cent. ____________

19. A supplier provides 1/2-inch elbows at a price of $.75 less 8% less 5%. What is the price for the elbows? Round values to the nearer whole cent. ____________

20. A 3-ton air conditioner condensing unit lists for $1 200. The supplier sells it to an installer at an 11% discount. The installer sells it to a homeowner at a 3% discount.
    a. What is the price the installer paid for the condensing unit?
    b. What is the price the homeowner paid?
    c. How much did the installer make on the deal?

a. ______________
b. ______________
c. ______________

Ratio and Proportion

Unit 19 RATIO

BASIC PRINCIPLES

- Review and apply the principles of ratios to the problems in this unit.
- Review denominate numbers in section 1 of the appendix.

Ratios are a way of comparing two numbers. When written as a mathematical expression, the ratio is written as two numbers separated by a colon (:). When written as a statement, a ratio is expressed as a ratio of one number **to** a second number. When written as a fraction, the first number of the ratio is the numerator of the fraction and the second number is the denominator.

In most cases, the ratio is a comparison of two whole numbers. So if the numbers being compared have fractions in them, an equivalent ratio is formed that has only whole numbers in it. Therefore you would not see a ratio of 2 1/2 to 3, but you would see 5 to 6. (This is the same ratio and was found by multiplying both numbers by 2.) You would also not see 2.5 to 3, but would see 5 to 6.

A ratio can always be set up as a fraction. The ratio is usually stated as value A to value B. The quantity after the word "to" always becomes the denominator of the fraction.

PRACTICAL PROBLEMS

Express each ratio as a fraction in lowest terms.

1. 4:8 ____________
2. 24:40 ____________
3. 15:3 ____________
4. 5/8:3/8 ____________
5. 12/13 to 8/13 ____________
6. 78 to 52 ____________

Find the inverse ratio of each. Express each answer as a fraction in lowest terms.

7. 7:3 ____________
8. 6/7:2/7 ____________
9. 12 to 16 ____________
10. 21 to 7 ____________

Note: Use this diagram for problems 11–18.

A

B

C

D

Measure each line to the nearer 1/8 inch. Using the measured lengths, find each ratio. Express each answer in lowest terms.

11. A:B ______
12. A:C ______
13. B:D ______
14. C:D ______
15. D:A ______
16. C:B ______
17. C:A ______
18. D:B ______
19. Refrigerant R-502 is a mixture of refrigerants R-22 and R-115. It takes 48.8 pounds of R-22 and 51.2 pounds of R-115 to make 100 pounds of R-502. Find the ratio of R-22 to R-115. ______

Note: Use this information for problems 20 and 21.
When two pulleys with different diameters are connected by a belt, the revolutions per minute for each pulley are different. The ratio of the revolutions per minute is the inverse of the ratio of the pulley diameters.

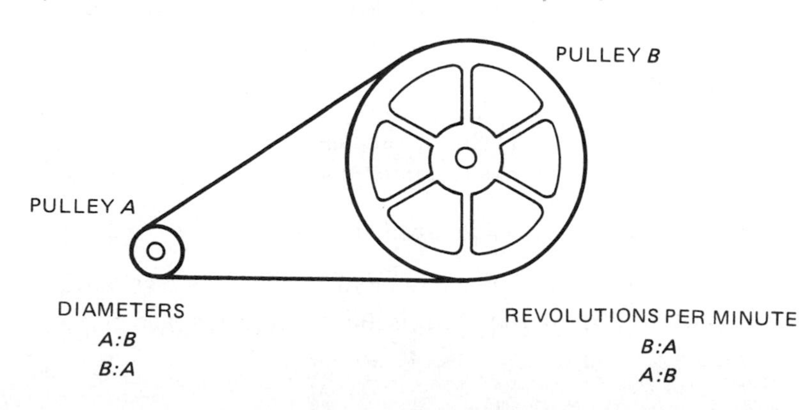

20. Find the ratio of the revolutions per minute for the fan pulley to the revolutions per minute for the motor pulley. __________

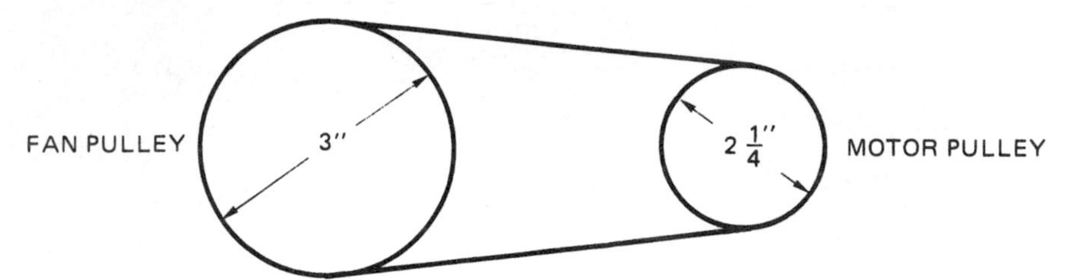

21. What is the ratio of the revolutions per minute for the motor to the revolutions per minute for the compressor? __________

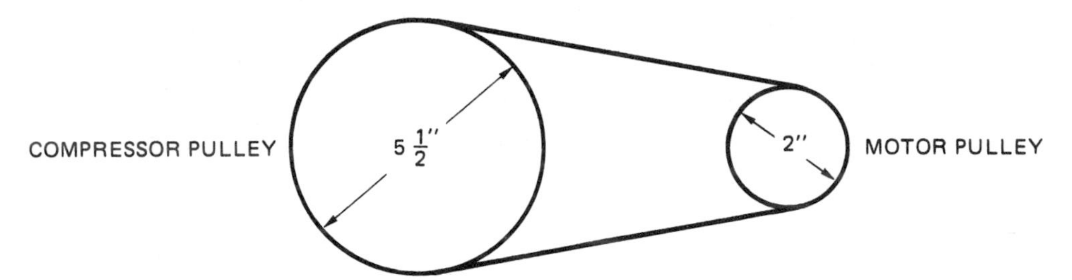

22. The weight of 1 cubic foot of Grade 2 fuel is about 53.125 pounds. The weight of 1 cubic foot of water is about 62.5 pounds. Find the ratio of the weight of the fuel to the weight of the water. __________

23. Two new air conditioning units are charged for the first time. To be fully charged, unit *A* needs 20 ounces of refrigerant. Unit *B* needs 14 ounces. Find the ratio of the amount of refrigerant used in unit *A* to the amount used in unit *B*. __________

Note: Use this information for problems 24 and 25.
The compression ratio of a compressor is given as the pressure of the refrigerant at discharge from the cylinder divided by the pressure of the refrigerant at inlet into the cylinder. Pressure is measured in pounds per square inch absolute (psia).

$$\text{Compression Ratio} = \frac{\text{Pressure at discharge (in psia)}}{\text{Pressure at inlet (in psia)}}$$

The ratio is then expressed as a decimal fraction.

24. A compressor takes in refrigerant at a pressure of 80.34 psia. The discharge pressure of the refrigerant is 281.74 psia. What is the compression ratio of the compressor? Round the answer to the nearer tenth. __________

25. A compressor takes in refrigerant R-12 at a pressure of 19.7 psia and compresses it to 154.7 psia. Find the compression ratio of the compressor. Round the answer to the nearer hundredth. ____________

26. In one minute, 90 cubic feet of air flow through a duct into a room. The room contains 960 cubic feet of space. What is the ratio of the flow of air into the room to the volume of the room? ____________

27. The intake of a compressor is 15 psig (pounds per square inch gauge reading). The exhaust has a pressure of 140 psig. Find the ratio of the exhaust pressure to the intake pressure. ____________

28. An installer compares the area of circular ducts and rectangular ducts. The 6-inch by 4-inch rectangular duct has an area of 24 square inches. A 5-inch diameter circular duct has an area of about 20 square inches. What is the ratio of the area of the circular duct to the area of the rectangular duct? ____________

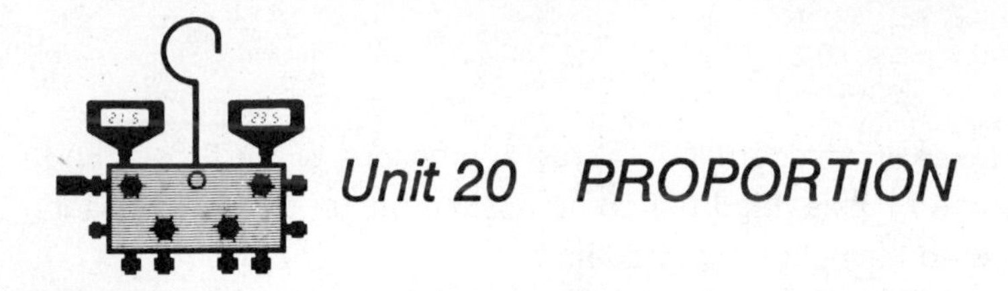

Unit 20 PROPORTION

BASIC PRINCIPLES

- Review and apply the principles of proportions to the problems in this unit.
- Review denominate numbers in section I of the appendix.

Proportions are the equivalence of two ratios. They can be set up by setting the two ratios equal to each other. Often one of the values is not known. This is found by solving the proportion.

$$\frac{A}{B} = \frac{C}{D}$$

This can be solved as

$$A = \frac{B \times C}{D}$$

if A is the quantity that is unknown, or

$$B = \frac{D \times A}{C}$$

if B is the unknown quantity.

If either C or D is unknown, switch the two fractions around so that the unknown is either A or B and then solve it as described above. All proportions can be solved in this manner.

Proportions are just 2 ratios set equal to each other. Set up the proportion as (different letters are used as numbers)

$$\frac{A}{B} = \frac{C}{D}$$

This can always be solved by cross multiplying

$$A \times D = B \times C$$

Then, to find the unknown quantity, divide the 2 numbers being multiplied together by the number multiplying the value you do not know.

PRACTICAL PROBLEMS

Find each unknown value.

1. $5/9 = x/27$ __________
2. $3/7 = 21/x$ __________

3. $7/21 = 6/x$ __________

4. $2/5 = x/17$ __________

5. 2:9 = 6:? __________

6. 11:3 = ?:15 __________

7. 3:5 = 7:? __________

8. 10:2 = ?:7 __________

9. The weight of 10 gallons of Grade 2 fuel oil is 71 pounds. What is the weight of 225 gallons? __________

10. The weight of 4 feet of 1-inch copper tubing is 2.62 pounds. Find the weight of 20 feet of the tubing. __________

Note: Use this information for problems 11 and 12.

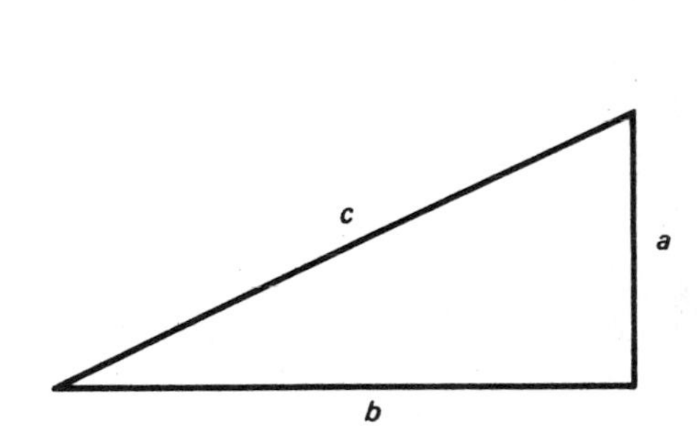

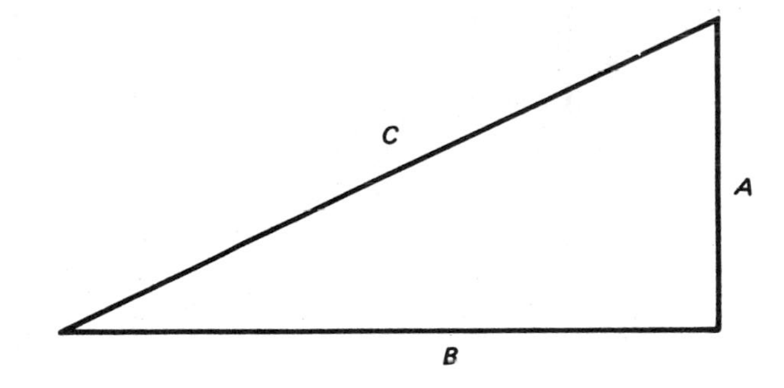

Corresponding sides of similar triangles (triangles which have equal angles) are in proportion. That is:

$$a:A = b:B$$
$$a:A = c:C$$
$$b:B = c:C$$

11. Measurements on this **Y** duct are taken from the center of the duct. Find dimension **X**. ____________

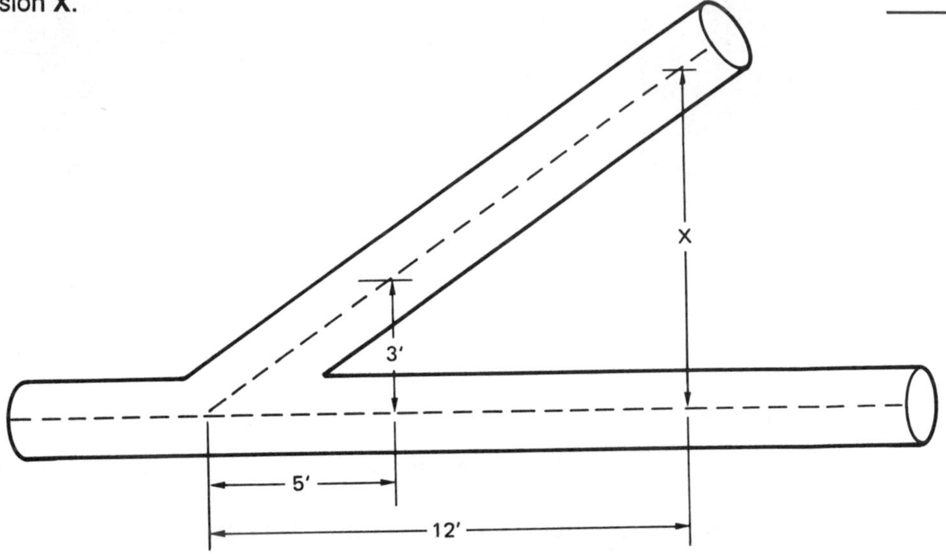

12. A fuel line from the tanks to the oil burner goes through the wall. Find dimension **Y.** ____________

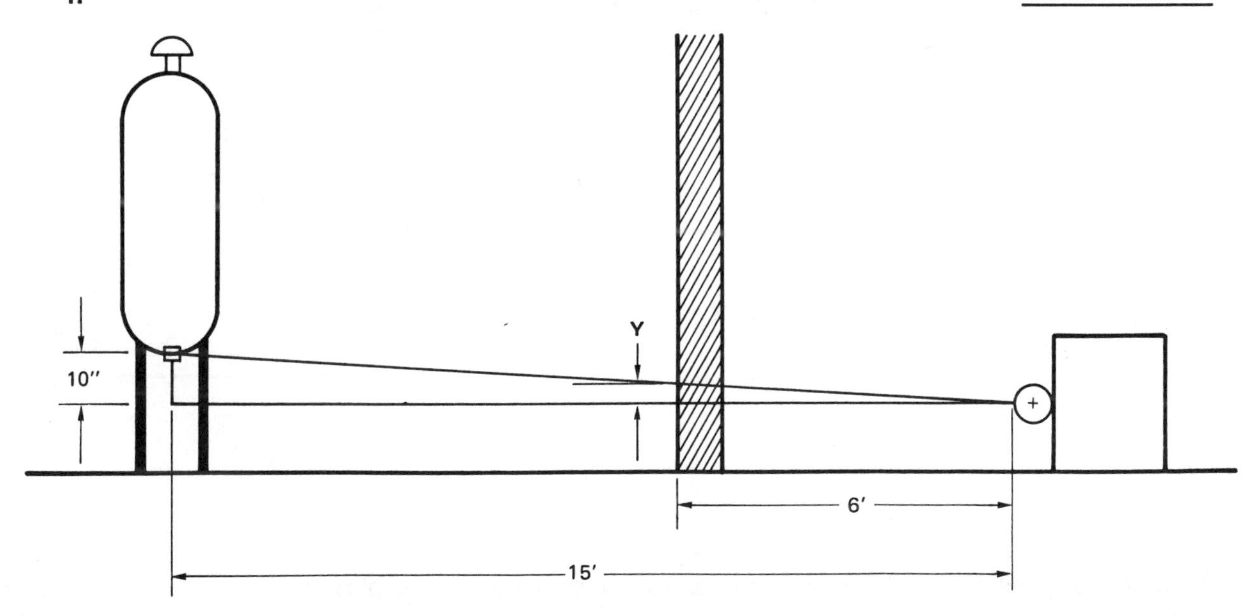

13. A compressor runs at a speed of 1 800 revolutions each minute. On each revolution, 7.2 cubic inches of gas enter the compressor. What is the volume of gas that enters the compressor each minute when it is running? ____________

14. The space in a cylinder when the piston is at the top of the stroke is the compressed volume of a compressor. A certain compressor has a compressed volume of 0.9 cubic inch. This means that for each revolution, the volume of compressed gas leaving the compressor is 0.9 cubic inch. The compressor makes 1 800 revolutions each minute. What is the volume of compressed gas leaving the compressor in each minute? __________

15. For each 4 feet of duct, the pressure in the duct drops 0.003 inch of water. What is the pressure drop between the ends of this duct? __________

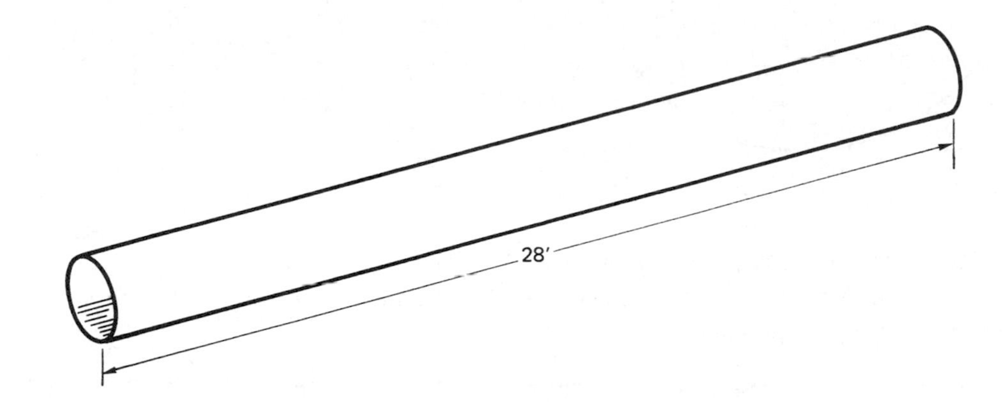

16. This motor pulley has a diameter of 2 inches and runs at 1 400 revolutions per minute (1 400 rpm). The fan pulley has a diameter of 3 1/2 inches. At how many revolutions per minute does the fan pulley revolve? __________

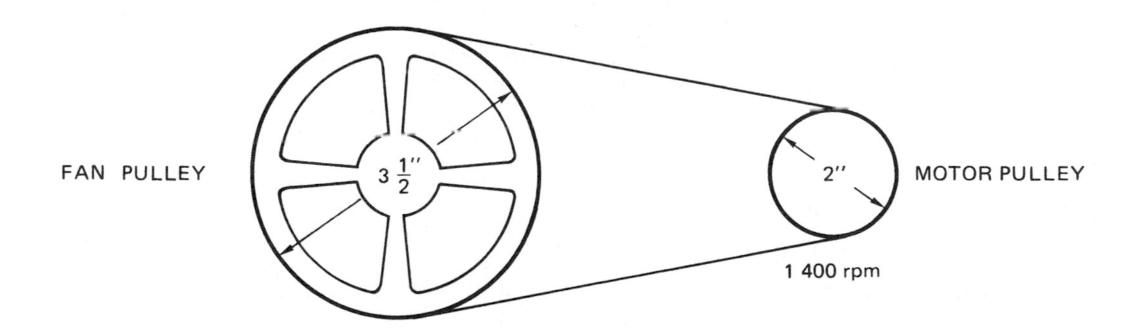

17. A compressor is run by a motor. If the compressor runs at 500 revolutions per minute (500 rpm), at how many revolutions per minute does the motor run? ______________

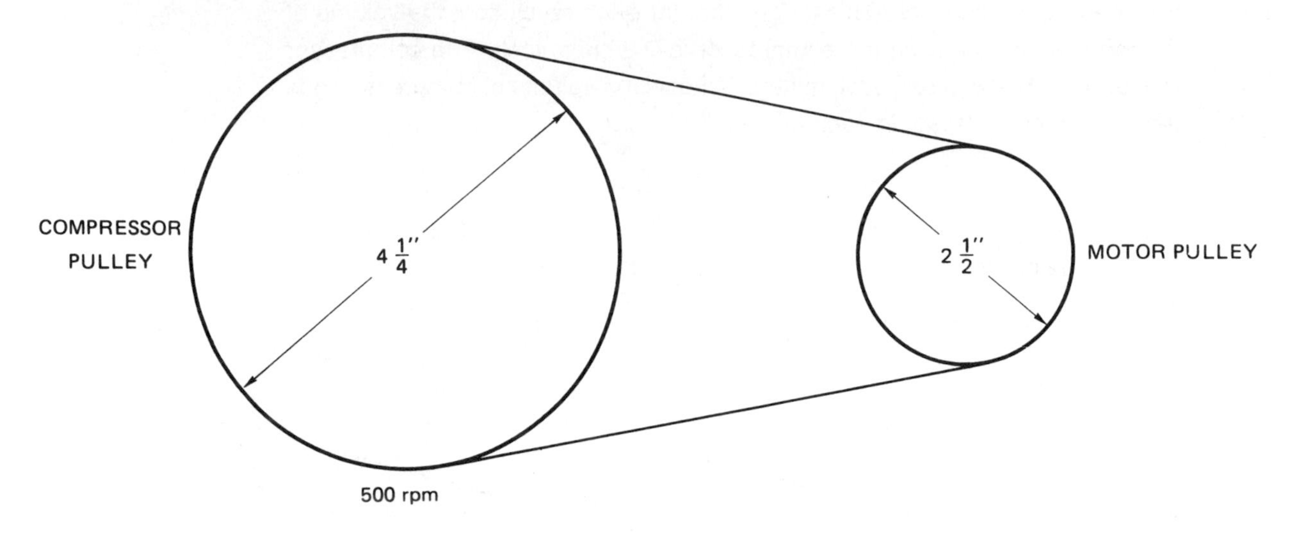

18. This blueprint shows the layout for rectangular ducting for a home forced air heating system. The dimensions are to be measured along the center of the ducts. Find, in feet, each dimension.

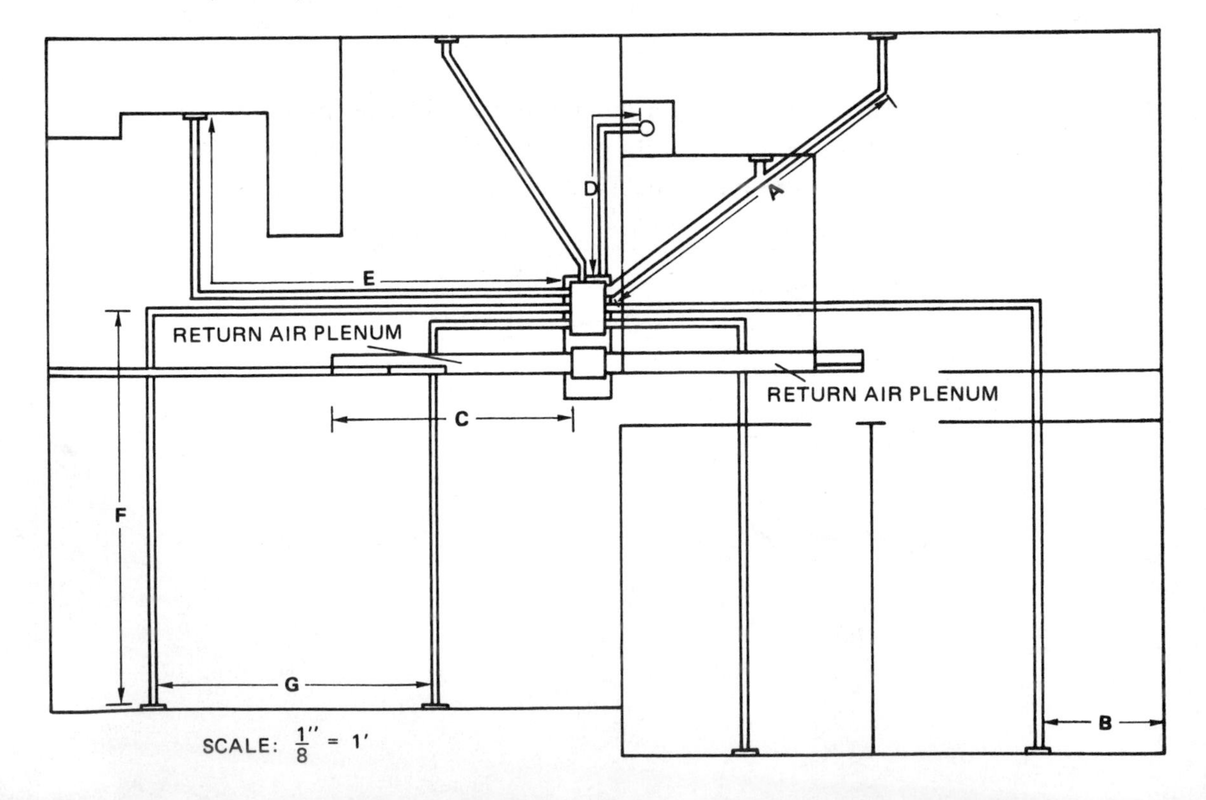

a. Dimension A
b. Dimension B
c. Dimension C
d. Dimension D
e. Dimension E
f. Dimension F
g. Dimension G

a. ______
b. ______
c. ______
d. ______
e. ______
g. ______
f. ______

19. It has been determined that a 768-square-foot house would require a 5-ton air conditioner. For this same region, what size unit would a 1 080-square-foot house require? Round off to the next higher half-ton value. ______

20. A 10 000 Btu window air conditioner requires 12 ounces of R-21 refrigerant to completely recharge the system. How much R-21 would be needed to refill a proportionally larger 12 000 Btu window air conditioner? ______

Direct Measure

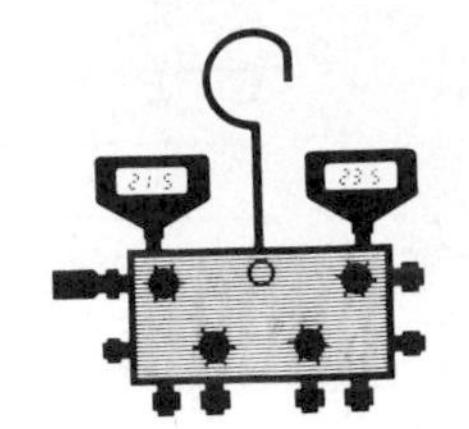

Unit 21 UNITS OF LENGTH MEASURE

BASIC PRINCIPLES

- Review and apply the principles of length measure to the problems in this unit.
- Review denominate numbers in section I of the appendix.
- Review equivalents in section II of the appendix.
- Study tables of units of length measure.

Often when solving problems involving units, all of the units must be the same. There are many times when the units in the problem are different, so they must be converted. There is a method that can always be used to convert from one unit to another correctly.

Suppose that we want to convert from 24 inches to feet. We all know that there are 12 inches to 1 foot, so 12 is the conversion factor; but do we multiply or divide by the 12? The following method will tell you which to do each time.

Write

$$24 \text{ inches}$$

Next to this write a fraction containing the units you are using so that the units you are starting with will get cancelled. In this case

$$24 \text{ inches} \left(\frac{\text{feet}}{\text{inches}} \right)$$

The inches will cancel out of the fractions leaving just feet as the final unit. Now fill in the fraction with numbers so that the numerator and denominator are equivalent. We know that 1 foot = 12 inches, so

$$24 \text{ inches} \left(\frac{1 \text{ foot}}{12 \text{ inches}} \right)$$

This is then solved just as a fraction problem. So the answer becomes

$$24 \text{ inches}\left(\frac{1 \text{ foot}}{12 \text{ inches}}\right) = 2 \text{ feet}$$

Getting to the unit that you want will often involve more than one conversion since you do not always know the direct conversion factor. The conversion can still be done; however, it may take 2 or more fractions.

Convert 15 840 inches to miles. We do not know the number of inches in a mile; however, we do know the number of inches in a foot and the number of feet in a mile, so we can make 2 conversions instead of 1.

$$15\,840 \text{ inches}\left(\frac{\text{feet}}{\text{inches}}\right)\left(\frac{\text{miles}}{\text{feet}}\right)$$

Once the units are straight, fill in the numbers. Remember that each fraction must be an equivalent, regardless of other fractions or numbers around.

$$15\,840 \text{ inches}\left(\frac{1 \text{ foot}}{12 \text{ inches}}\right)\left(\frac{1 \text{ mile}}{5280 \text{ feet}}\right) = 0.25 \text{ miles or } \frac{1}{4} \text{ mile}$$

There are times when numerators other than 1 get multiplied together. It depends upon the conversion that must be made.

Tables of some unit equivalents are included below.

ENGLISH LENGTH MEASURE		
1 foot (ft)	=	12 inches (in)
1 yard (yd)	=	3 feet (ft)
1 mile (mi)	=	1 760 yards (yd)
1 mile (mi)	=	5 280 feet (ft)

METRIC LENGTH MEASURE		
10 millimeters (mm)	=	1 centimeter (cm)
100 centimeters (cm)	=	1 meter (m)
1 000 meters (m)	=	1 kilometer (km)

To convert units correctly time after time, do the same process each time. An easy way to do this is to treat the problem as the original number and unit being multiplied by a fraction. The denominator of the fraction contains the original unit. The numerator has the unit the answer should have. The numbers that go with these units usually come from a chart that has these units as equivalents. Setting the problem up this way tells you whether to multiply or divide to get the answer. A calculator can be used to help multiply or divide the numbers.

As an example: 3 feet = ? inches

$$3 \text{ feet} \times \frac{\text{inches}}{\text{feet}}$$

$$3 \text{ feet} \times \frac{12 \text{ inches}}{1 \text{ foot}}$$

$$3 \text{ feet} \times \frac{12 \text{ inches}}{1 \text{ foot}} = 36 \text{ inches}$$

PRACTICAL PROBLEMS

Express each measurement in inches.

1. 5 feet ______
2. 4 1/2 feet ______
3. 7 1/6 feet ______
4. 3.4 feet ______

Express each measurement in centimeters.

5. 1.75 meters ______
6. 3 meters ______
7. 0.07 meter
8. 2.505 meters ______

Express each measurement in feet.

9. 60 inches ______
10. 108 inches ______
11. 50 inches ______
12. 42 inches ______

Express each measurement in meters.

13. 465 centimeters ______
14. 29 centimeters ______
15. 550 centimeters ______
16. 0.9 centimeter ______
17. Express 2 feet 7 inches as inches. ______
18. Express 6 feet 10 inches as feet. ______
19. Express 3 feet 9 inches as inches. ______
20. Express 8 feet 4 inches as feet. ______

21. Express the dimensions of each room in feet.

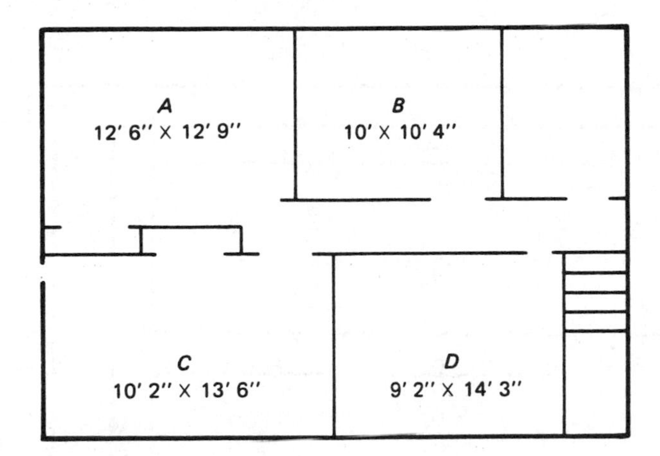

a. Room *A* a. ____________

b. Room *B* b. ____________

c. Room *C* c. ____________

d. Room *D* d. ____________

22. The studs in a wall are spaced 1 foot 4 inches apart (center to center). The holes in an electric baseboard heater need to be how many inches apart so that each screw goes into a stud?

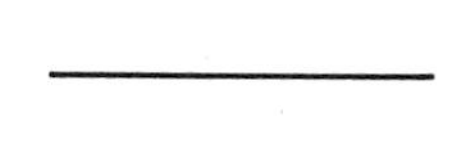

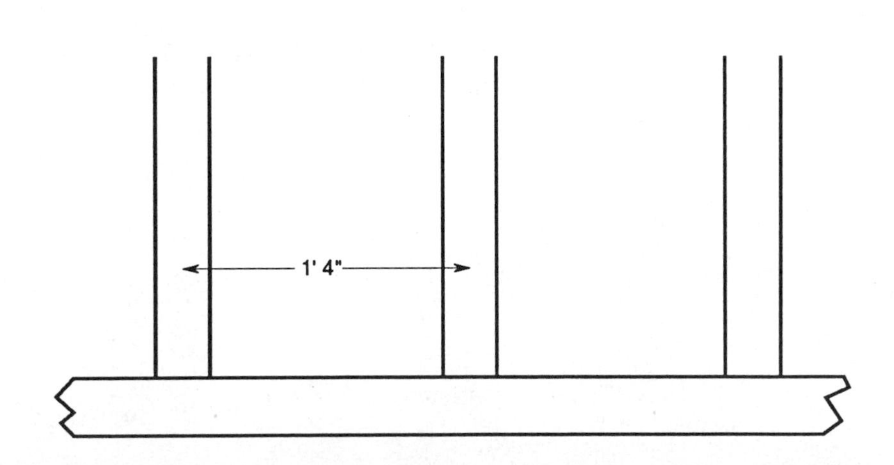

23. A window is 2 feet 7 inches across. What is the largest width air conditioner in inches that will fit in that window? ______

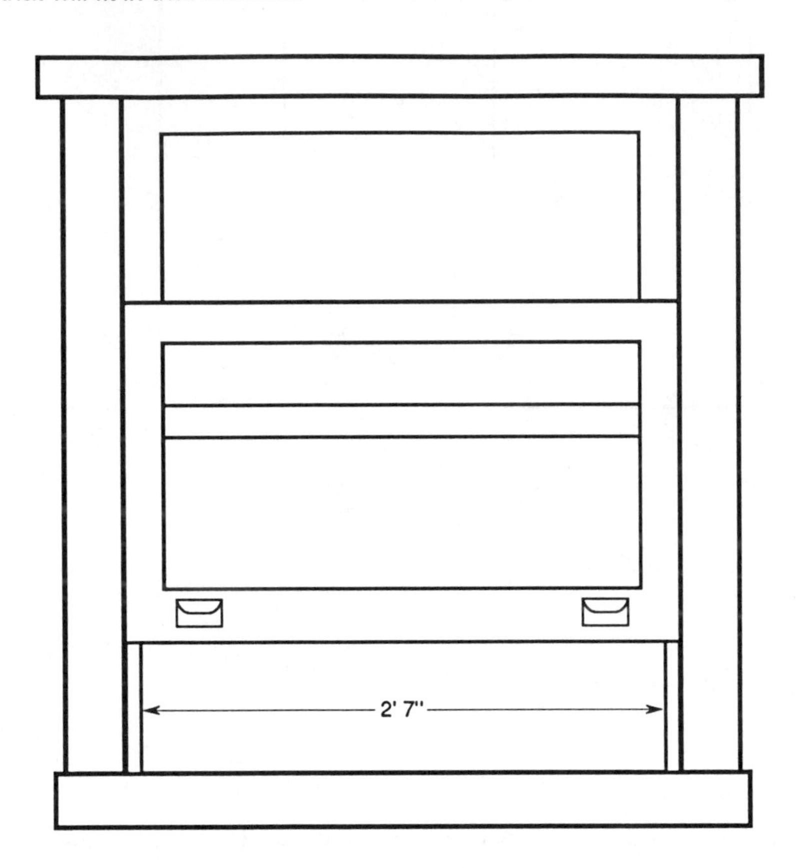

24. A strap to support a round duct is 1.27 meters long. Find the length of the strap in centimeters. ______

25. A domestic heat pump system has the condenser coils and evaporator coils separated by 10.46 meters. What would be the length of the hose expressed in centimeters, connecting these 2 coils? ______

Unit 22 EQUIVALENT UNITS OF LENGTH MEASURE

BASIC PRINCIPLES

- Review and apply the principles of equivalent units of length measure to the problems in this unit.
- Review denominate numbers in section I of the appendix.
- Review equivalents in section II of the appendix.
- Study these tables of equivalent units of length measure.

With many items now being supplied by foreign manufacturers, many instructions are including metric measurements. It is important to be able to quickly and correctly convert from one unit system to the other.

In the last unit a method was presented to convert between units of measure. This method works whether in the English unit system, the metric unit system, or both systems. In the case of converting between unit sytem, one of the fractions has to be the equivalent between English and metric units. The method is the same as before. As a matter of fact, this method will work when converting units of length, weight, area, volume, or any type of unit that you want. This one method will do it all!

A table of equivalences between English and metric units is given below.

1 millimeter (mm)	=	0.039 37 inch (in)
1 centimeter (cm)	=	0.393 7 inch
1 meter (m)	=	39.37 inches (in)

1 inch	=	25.4 millimeters (mm)
1 inch	=	2.54 centimeters (cm)
1 inch	=	0.025 4 meter (m)

1 foot (ft)	=	0.304 8 meter (m)
1 yard (yd)	=	0.914 4 meter (m)
1 mile (mi)	=	1.609 kilometers (km)
1 meter (m)	=	3.280 84 feet (ft)
1 meter (m)	=	1.093 61 yards (yd)
1 kilometer (km)	=	0.621 37 mile (mi)

PRACTICAL PROBLEMS

Note: For problems 1–6, round to the nearer hundredth when needed.

1. Express 9 inches as centimeters. ____________
2. Express 2 feet as centimeters. ____________
3. Express 1 foot 7 inches as centimeters. ____________
4. Express 3 feet as meters. ____________
5. Express 7 feet as meters. ____________
6. Express 5 feet 8 inches as meters. ____________

Note: For problems 7–12, round to the nearer thousandth when needed.

7. Express 15 centimeters as inches. ____________
8. Express 2 meters as inches. ____________
9. Express 86 centimeters as feet and inches. ____________
10. Express 4 meters as feet. ____________
11. Express 7 meters as feet and inches. ____________
12. Express 6.3 meters as feet and inches. ____________
13. Find, in centimeters, the diameter of this duct. ____________

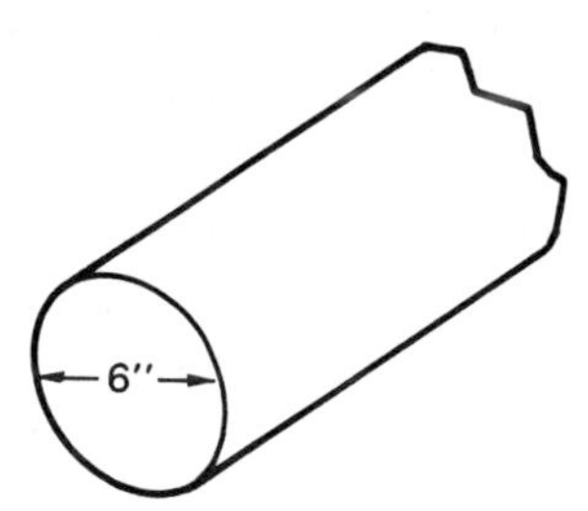

14. The connecting rod in a compressor for an air conditioning system is 3.5 inches long. How long is the rod in centimeters? ____________

15. The dimensions of this room are in meters.

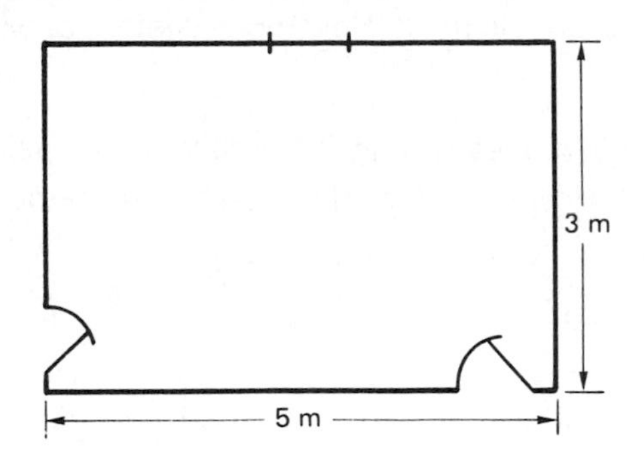

  a. Find the width of the room in feet and inches. Round the inches to the nearer tenth. a. ____________

  b. Find the length of the room in feet and inches. Round the inches to the nearer tenth. b. ____________

16. An anemometer measures the velocity of the air from a grill. The anemometer reading is 18 feet per minute. What is the reading in meters per minute? ____________

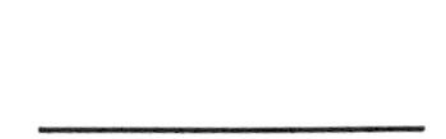

17. The screws on the this supply plenum are equally spaced. The center-to-center distance between screws is 25 centimeters. What is the distance in inches? ____________

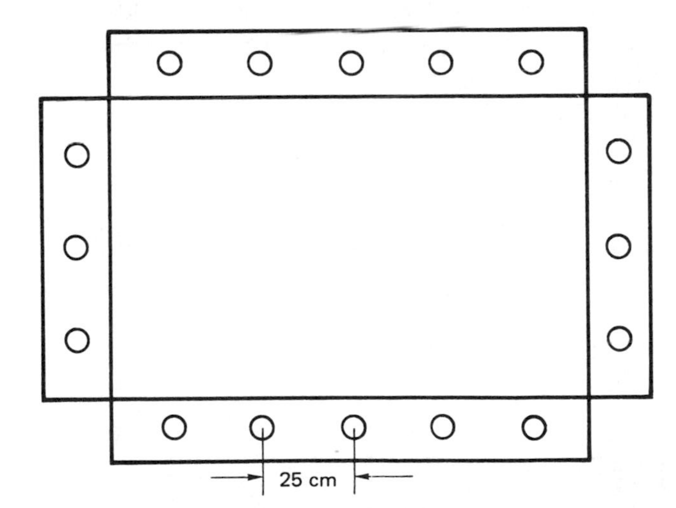

18. A section of tubing for a baseboard hot water heating system must be replaced. The section is 2 feet 8 inches long. The replacement is measured in centimeters. What is the length of the tubing in centimeters? ____________

19. To get power to a compressor for an air conditioner, 11 feet of wire are needed. A roll of wire with 15 meters of wire on it is used. How many meters of wire are left? ________

20. A refrigerator door needs a gasket 3.48 meters long. The roll that the gasket is to be taken from contains 25 feet. How many feet of gasket are left? Round the answer to the nearer tenth foot. ________

21. The dimensions of this duct are in feet.

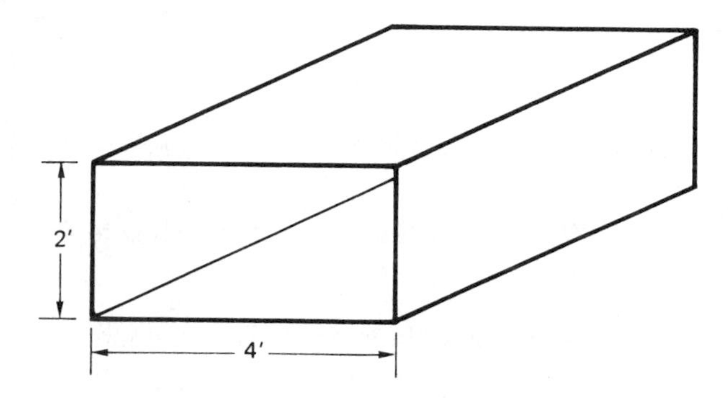

a. Find, in centimeters, the width of the duct. a. ________
b. Find, in centimeters, the height of the duct. b. ________

22. Flue pipe connects an oil furnace with the chimney of a house. The flue is 4 feet long and is inclined so that it rises 6 inches from the furnace to the chimney.
a. What is the length of the flue in meters? a. ________
b. What is the rise of the flue in centimeters? b. ________

23. A forced air system uses this filter.

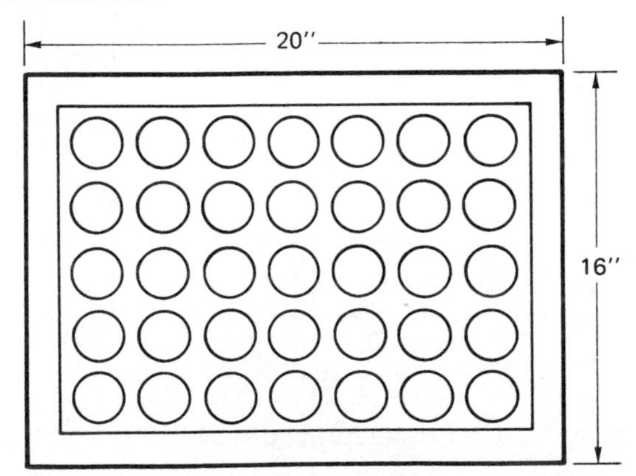

a. Find, in centimeters, the length of the filter. a. ________
b. Find, in centimeters, the width of the filter. b. ________

24. The dimensions of this gas furnace are in inches. Round each answer to the nearer tenth meter.

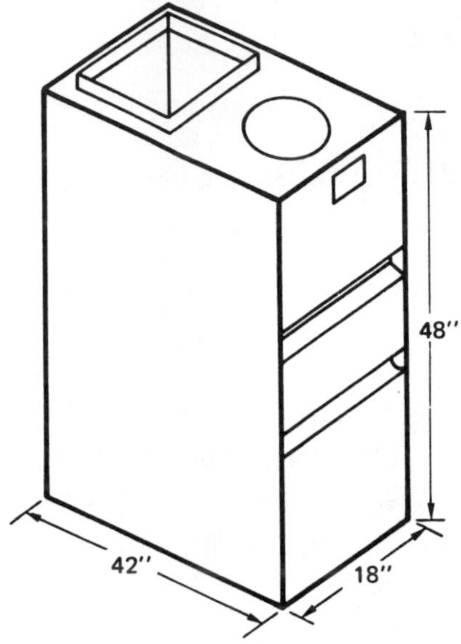

    a. What is the height of the burner? a. ____________
    b. What is the width of the burner? b. ____________
    c. What is the length of the burner? c. ____________

25. A condensing coil has 12 fins per inch. How many fins per centimeter does the coil have? Round the answer to the nearer hundredth. ____________

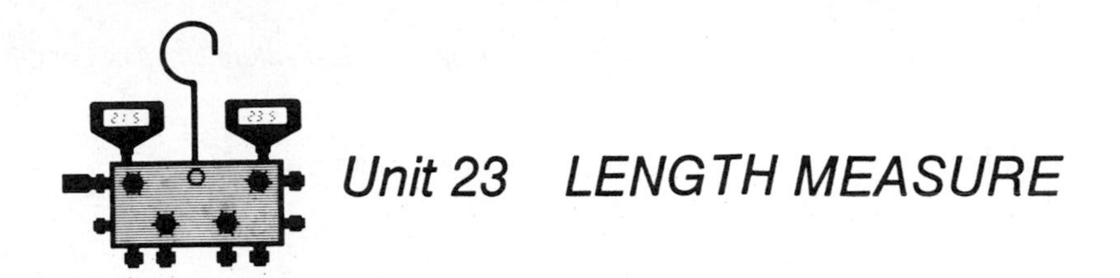

Unit 23 LENGTH MEASURE

BASIC PRINCIPLES

- Review and apply the principles of length measure to the problems in this unit.
- Review denominate numbers in section I of the appendix
- Review the tables of length measure in section II of the appendix.
- Use formulas for perimeters found in section III of the appendix.

Adding or subtracting lengths requires that all of the lengths be the same units. An exception to this might be when adding lengths in feet and inches; these could be added by converting all the measurements to feet or all to inches and then doing the addition; a second way to find the sum would be by adding all of the inches first. If the total is larger than 12 inches, convert to feet and inches; then add this feet value with the other feet values.

As an example, add:

3 feet 5 inches
4 feet 7 inches
8 feet 10 inches

First add 5 inches + 7 inches + 10 inches = 22 inches. Also, 3 feet + 4 feet + 8 feet = 15 feet. Now 22 inches is larger than 12 inches, so convert it: 22 inches = 1 foot 10 inches. Then add this 1 foot to the 15 feet to get 16 feet. The final problem and answer becomes:

3 feet 5 inches
4 feet 7 inches
8 feet 10 inches
15 feet 22 inches = 16 feet 10 inches

When subtracting, you can use a similar procedure by borrowing 1 foot and converting it to 12 inches before subtracting if necessary.

Most of the time, the dimensions will all be in one unit and these types of conversions will not have to be done. Make sure that all of the measurements that are being combined have the same units (with the exception given above).

PRACTICAL PROBLEMS

1. The duct shown must be wrapped with insulation. The width of the insulation is designed to wrap completely around the duct. Find, in feet and inches, the length of insulation to complete the job. ____________

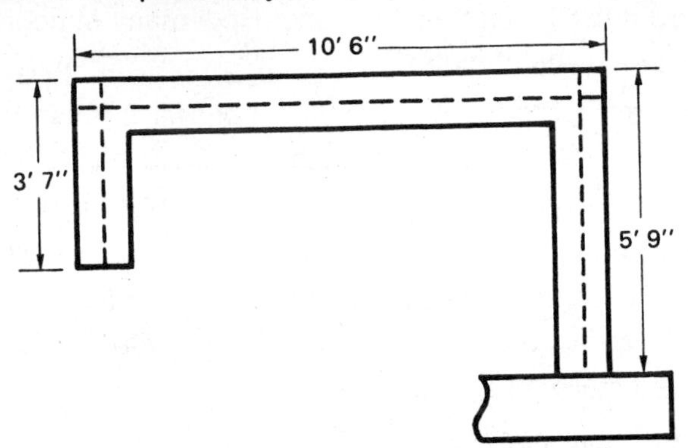

2. A piece of polyvinylchloride (PVC) drain pipe is used in an air conditioning unit. The pipe is 243 centimeters long. It is cut from a pipe which is 5 meters in length. How much pipe is left? ____________

3. A flue from an oil furnace to the chimney is made from 9 pieces of pipe. After being fitted together, each piece of pipe is 1 foot 10 1/2 inches long. What is the total length of the flue? ____________

4. Find, in meters, the length of straight duct used for this duct. ____________

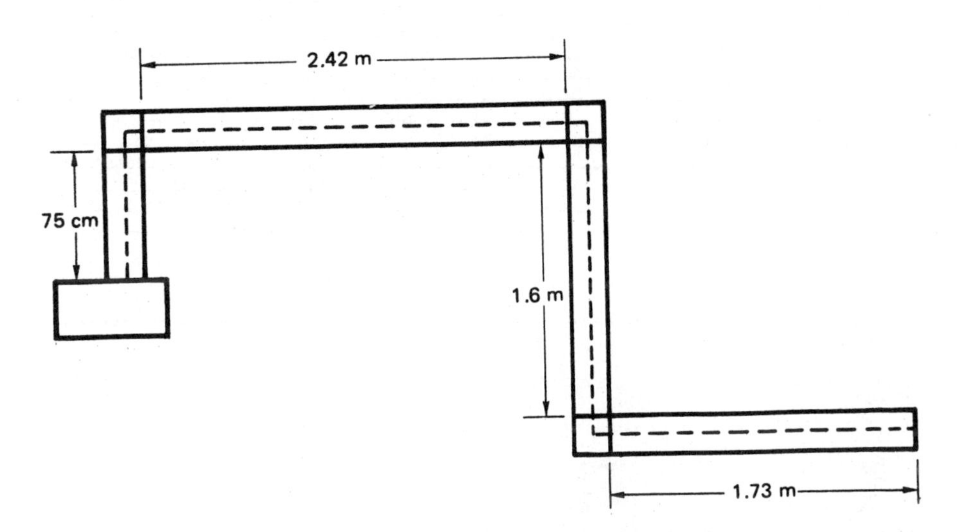

5. In repairing a condenser coil, a piece of tube 65 centimeters long is used.
   a. What is the length of tube needed for 16 pieces? Express the answer in centimeters.
   b. How many meters long are the 16 pieces of tube?

   a. ___________
   b. ___________

6. Support straps for a duct are each 1 foot 2 inches long. How many support straps can be cut from a 40-foot piece of metal?

   ____

7. The control wire from the thermostat to the oil burner control switch runs the following distances: 1 1/2 inches, 5 feet 4 1/2 inches, 3 inches, 5 1/2 inches, 12 feet 6 inches, 7 feet 2 1/4 inches, 8 inches. Find the total length of the control wire.

   ____

8. When installing an air conditioner in a car, hose with a diameter of 2.5 centimeters is used. Each end of each hose used must overlap 2 centimeters. The lengths needed, without overlap, are 3.24 meters and 3.47 meters. What is the total length of hose the installer must cut?

   ____

9. The openings in this grill are equally spaced.

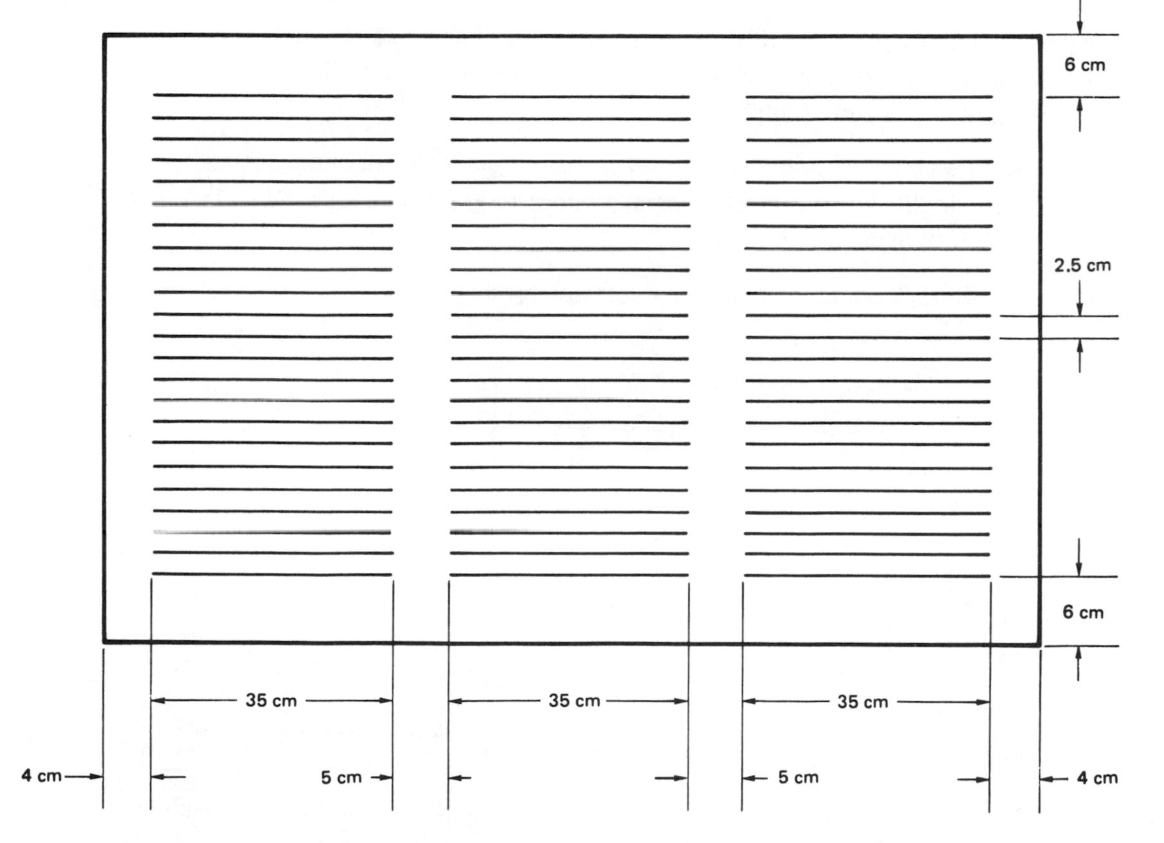

   a. Find, in meters, the width of the grill.
   b. Find, in meters, the height of the grill.

   a. ___________
   b. ___________

10. This electrical conduit is made from 5/8-inch diameter tubing joined to 5/8-inch diameter elbows. What is the total length of straight tubing used for the conduit? Express the answer in feet and inches. __________

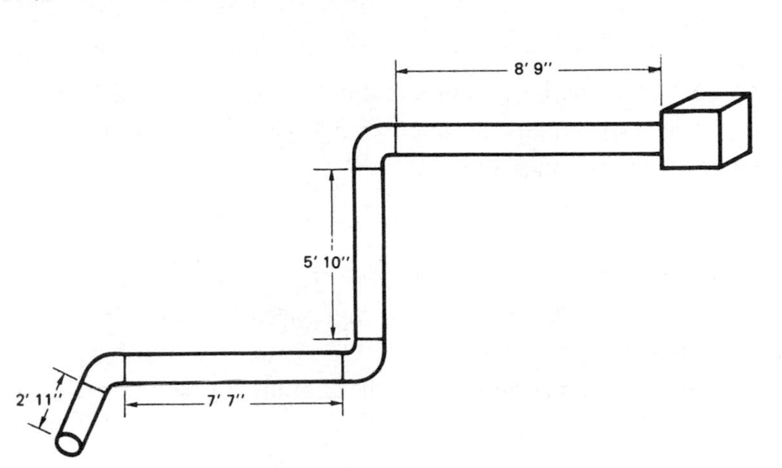

11. a. A bolt must pass through the layers shown. What is the thickness of the material to be bolted? Express the answer in centimeters. a. __________

    b. A 0.75 centimeter nut is to be screwed onto the bolt. The head of the bolt is 0.8 centimeter thick. What is the minimum length of the bolt to completely fill the nut? b. __________

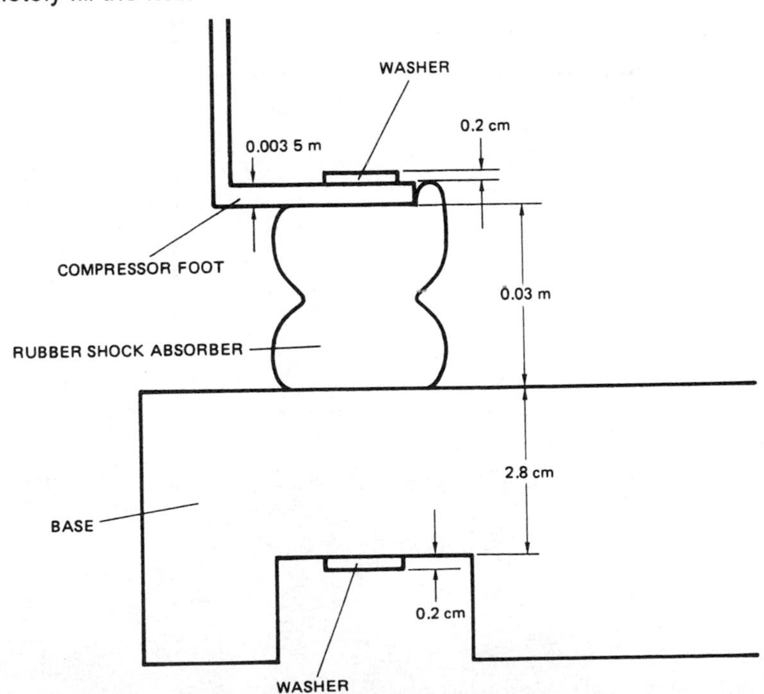

12. Pieces of copper tubing are used to install a hot water heating system. How many pieces, each 2 feet 4 inches long, can be cut from a 20-foot length of tubing? ____________

13. To repair a certain refrigerator, these lengths of wire are needed: 6 ft 4 in, 2 ft 3 in, 1 ft 11 in, and 4 ft 9 in. The lengths are cut from a 25-foot coil. Find, in feet and inches, the amount of coil left after the lengths are cut. ____________

14. Tape is used to seal insulation around 12 pieces of ducting. Each piece of duct needs two strips of tape. The first strip of tape measures 1.84 meters. The second strip measure 55 centimeters. The strips are cut from a 75-meter long roll of tape. How much tape will be left after the 12 pieces of ducting are sealed? ____________

Note: Use this information for problems 14–22.
The distance around any figure is called the *perimeter*.

The perimeter of a rectangle is:

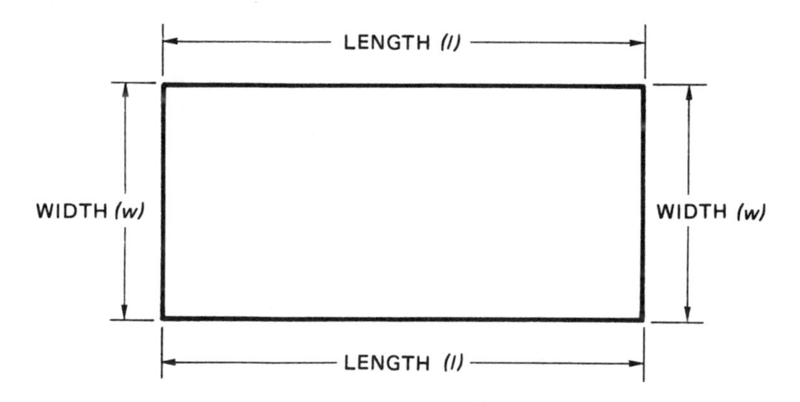

perimeter = length + width + length + width

or

$P = 2l + 2w$

The perimeter of a square is:

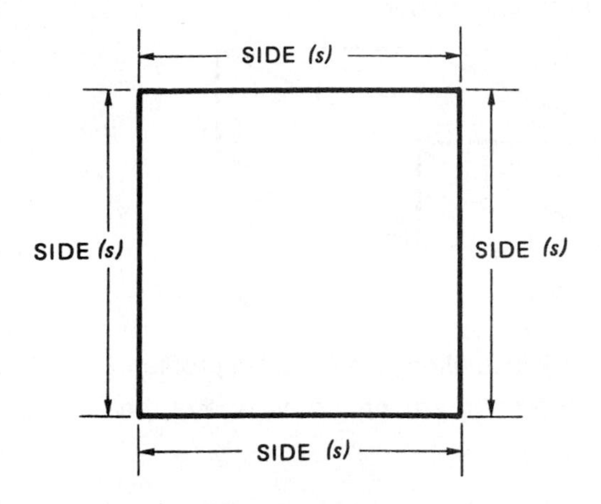

$$\text{perimeter} = \text{side} + \text{side} + \text{side} + \text{side}$$

or

$$P = 4s$$

The perimeter of a circle is called the *circumference*.

The circumference (C) is:

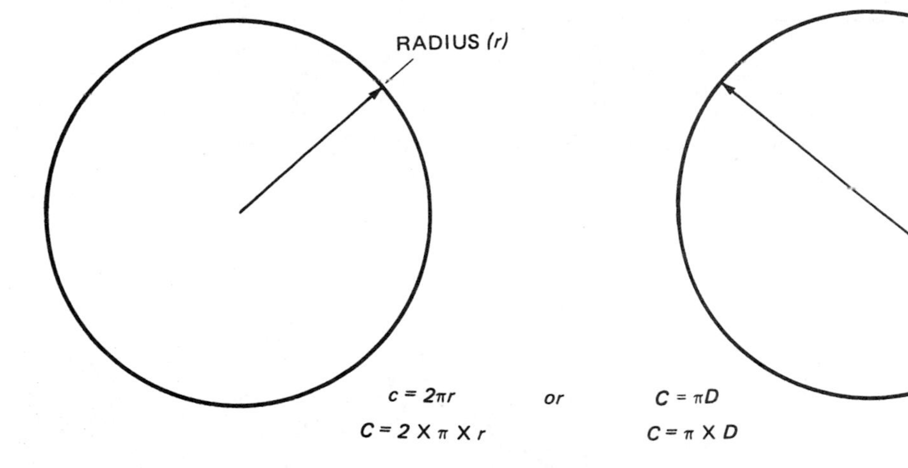

$c = 2\pi r$ *or* $C = \pi D$

$C = 2 \times \pi \times r$ $\quad$ $C = \pi \times D$

where $\pi = 3.141\ 6$

15. This rectangular duct is to be wrapped with insulation. How many inches of insulation are need to completely wrap the duct? ________

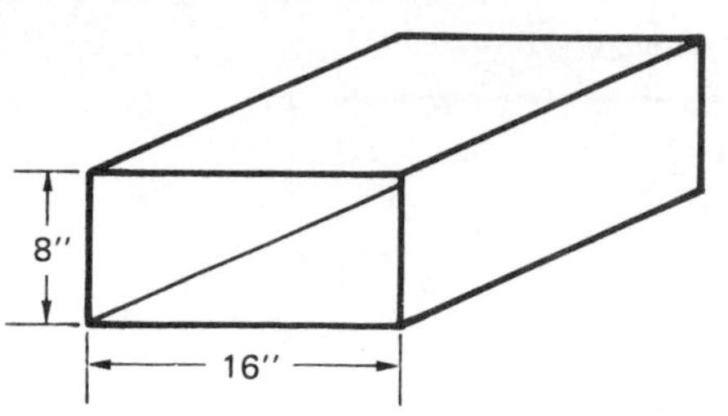

16. A band holds the 1 1/2 inches-thick insulation on the duct in problem 15. What is the minimum length of the band to go around the outside of the duct and have a 2-inch overlap? ________

17. A magnetic strip fits around the grill of this air conditioning unit. How long must the strip be to completely fit around the grill? ________

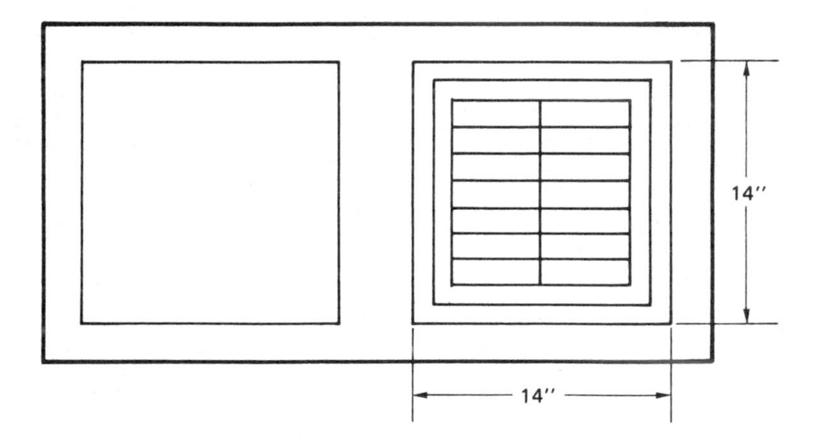

18. What length of insulation is needed to wrap this circular duct? Round the answer to the nearer hundredth inch. ________

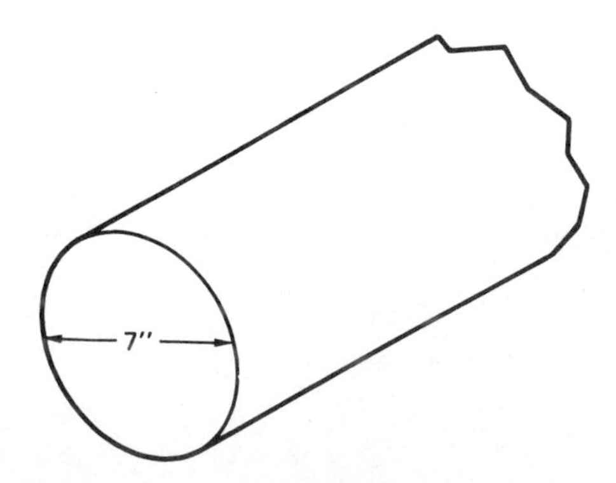

19. The thickness of the insulation in the above problem is 1 inch. What length of tape would be needed to wrap around the outside of the insulation. (Include a 2-inch overlap for the tape.) Round off the answer to the nearer hundredth inch. __________

20. A refrigerator door is sealed with a magnetic gasket. The rectangular door is 36 inches wide and 39 1/2 inches long. Find, in feet and inches, the total length of the gasket. __________

21. This is an end piece for the support frame of an A-type evaporator coil. To force the air flow through the coil, weather stripping is needed. The stripping is measured from the center of the strip. How many inches of stripping are needed? __________

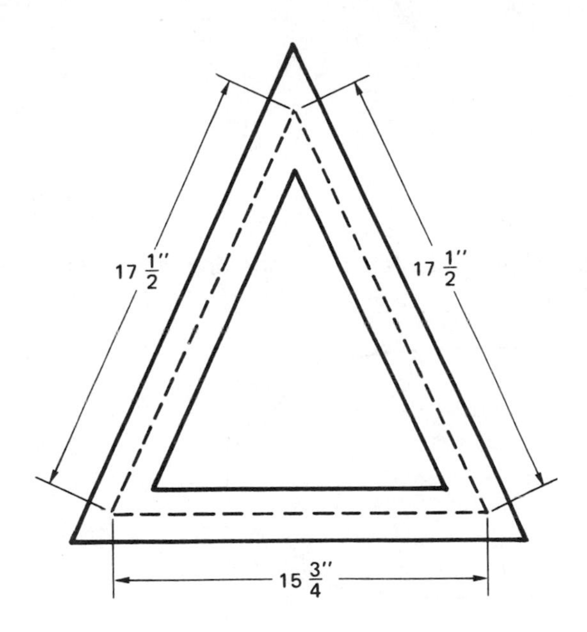

22. A compressor motor has a shaft diameter of 7/8 inch. The felt wiper which wraps tightly around the shaft must be replaced. What length of felt must be used? Round the answer to the nearer hundredth inch. __________

23. A room air conditioner is being installed in an opening in a wall. The air conditioner is 15 1/4 inches high and 26 3/4 inches long. Find the length of weather stripping needed to seal around the unit. __________

24. This bracket is needed to support a window air conditioning unit. What length of metal is needed to make the bracket? __________

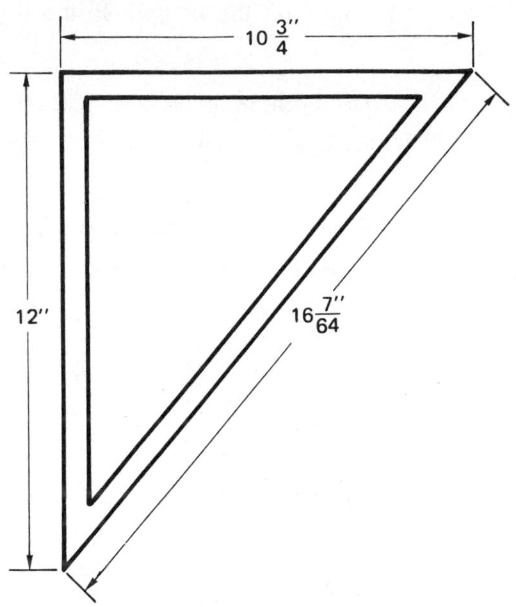

25. Find, to the nearer thousandth inch, the circumference of the ring which fits around this piston for a compressor. __________

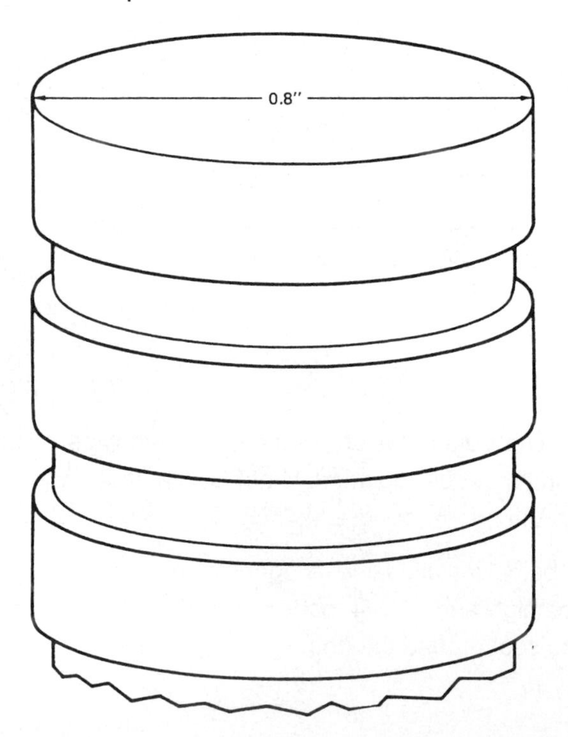

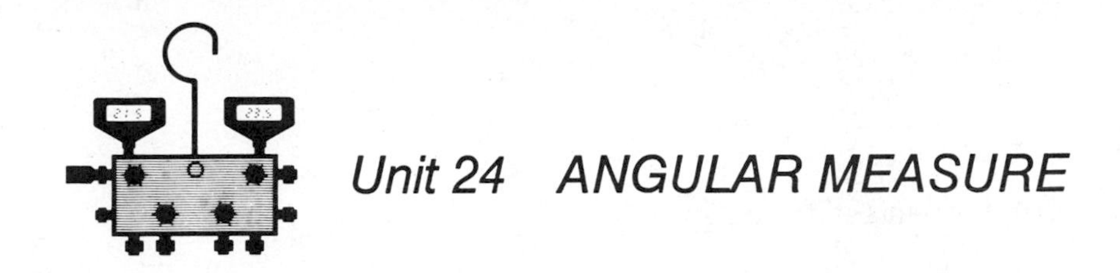

Unit 24 ANGULAR MEASURE

BASIC PRINCIPLES

- Review and apply the principles of angular measure to the problems in this unit.

This unit, unlike most of the others, has almost no calculations in it. It primarily involves simple measuring angles. Angles are measured with a protractor. If any calculations do have to be made, remember that a full circle has 360° in it.

A protractor is a half circle (semicircle) with degree markings on it. Although protractors come in different sizes and markings, all protractors do have some common points. As a rule, all protractors have degree markings for each degree. Sometimes the markings go from 0° to 180° in each direction, while others go from 0° to 90° to 0° again.

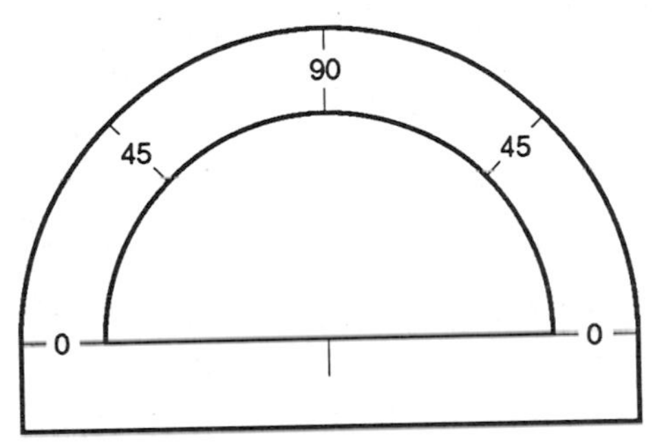

An angle is made when two straight lines meet. The two lines are known as sides, or legs, of the angle and they meet at the apex, or vertex, of the angle.

The size of an angle is measured by placing the center point of the protractor (the mark on the line between the two 0's) on the apex of the angle. Rotate the protractor so that the line to one of the 0's lies right on top of one of the sides of the angle while keeping the center point of the protractor on the apex of the angle. Next simply read off the angle where the other side of the angle crosses the protractor. If the angle is greater than 90°, care must be taken to read the correct scale or to count the number of degrees greater than 90. Many times the sides of the angle are small. These sides can be extended before measuring the angle.

If the angle is greater than 180°, measure the smaller angle and then subtract that angle from 360°.

Remember: To measure an angle using a protractor, always place the center mark of the flat edge at the apex or vertex of the angle (the point where the 2 lines meet). Rotate the protractor around that point so that one section of the flat edge is aligned with one of the lines (sides) of the angle. The size of the angle is read where the other side of the angle crosses the curved section of the protractor.

If the angle is larger than a straight line (180°), measure the smaller angle and subtract it from 360°.

PRACTICAL PROBLEMS

Measure each angle to the nearest degree.

1. 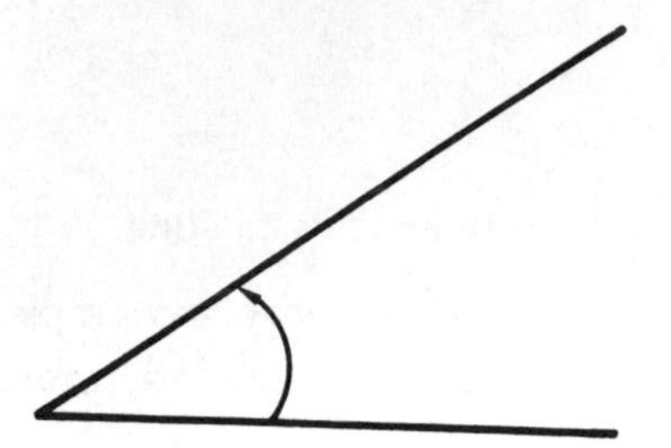

2. 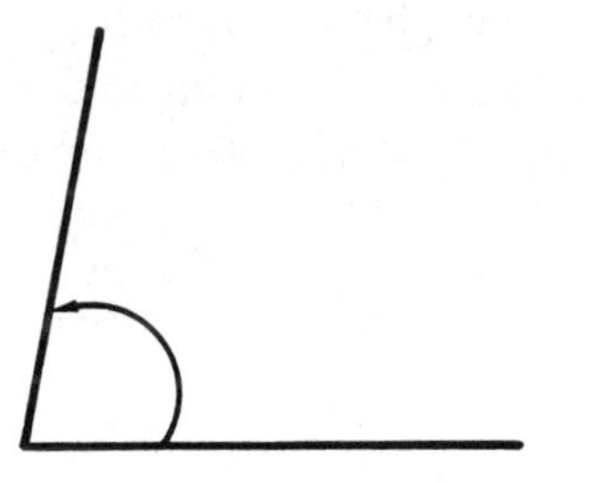

3. 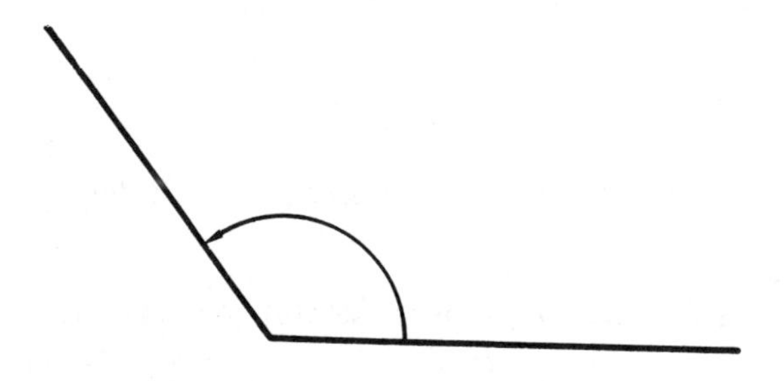

4. 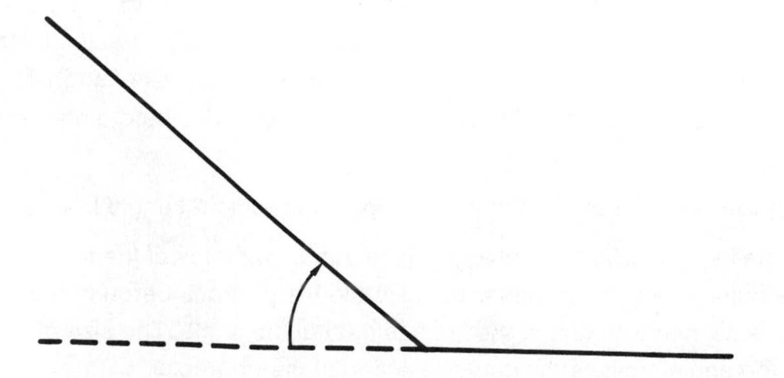

5. ____________

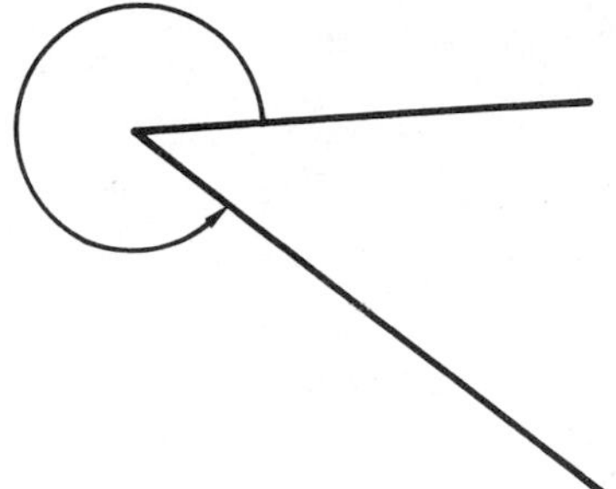

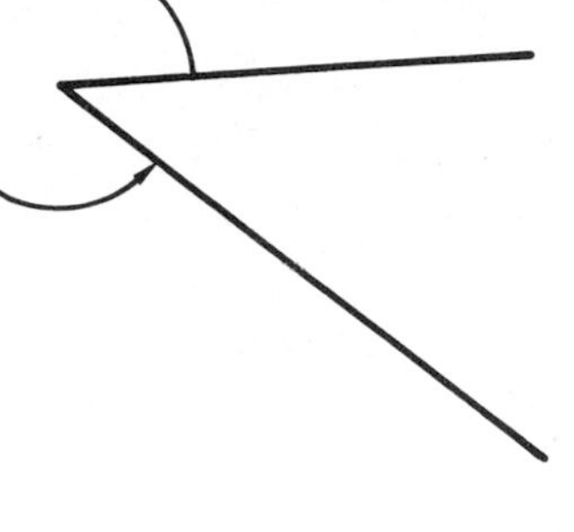

6. ____________

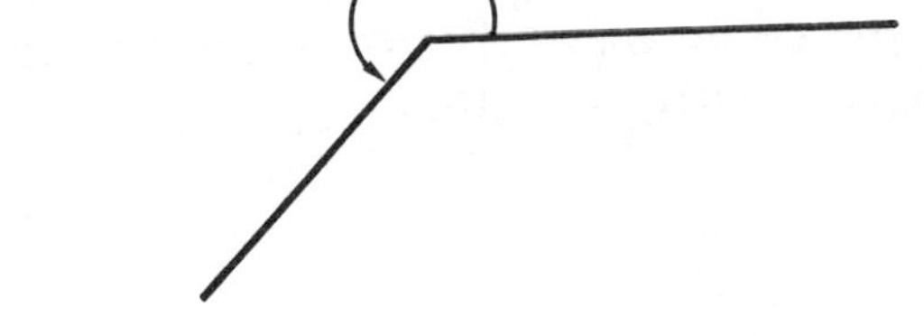

7. How many degrees are in the angle formed by the **Y** in this duct? ____________

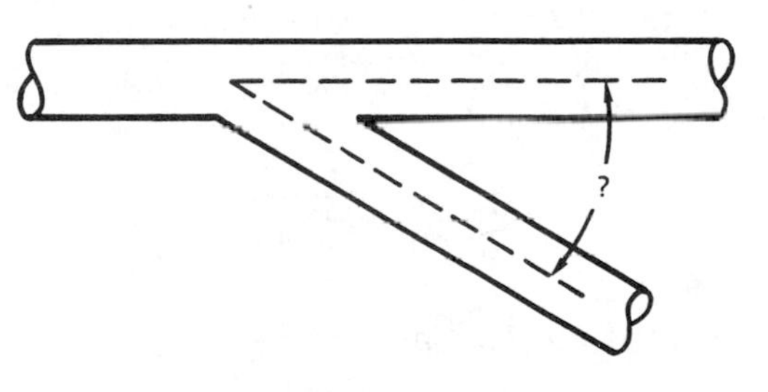

8. What is the angle made by the tip of this flaring cone? ____________

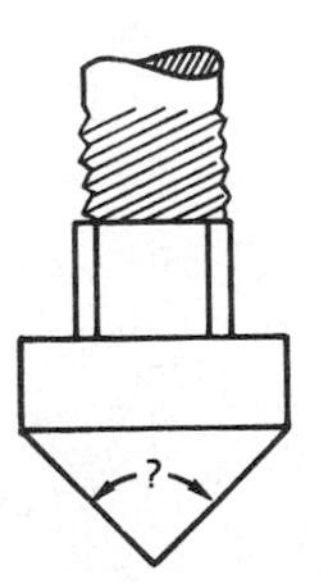

9. Find the number of degrees in the angle formed by the two parts of this A-type evaporator coil. __________

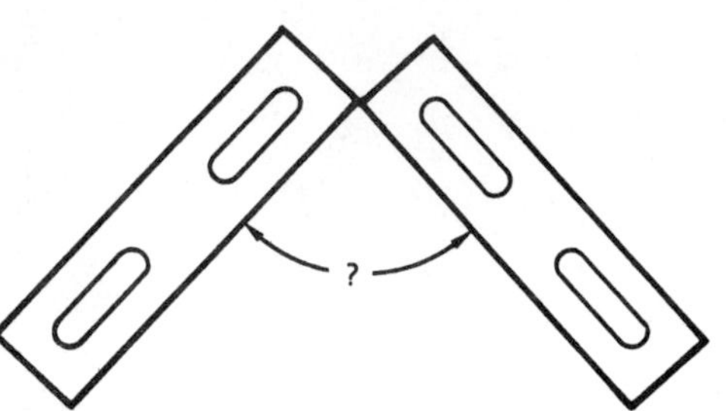

10. The control shown is for a window air conditioner.
    a. Through what angle must this dial be turned to go from Off to Low Cool? a. __________
    b. Through what angle must this dial be turned to go from Off to High Cool? b. __________

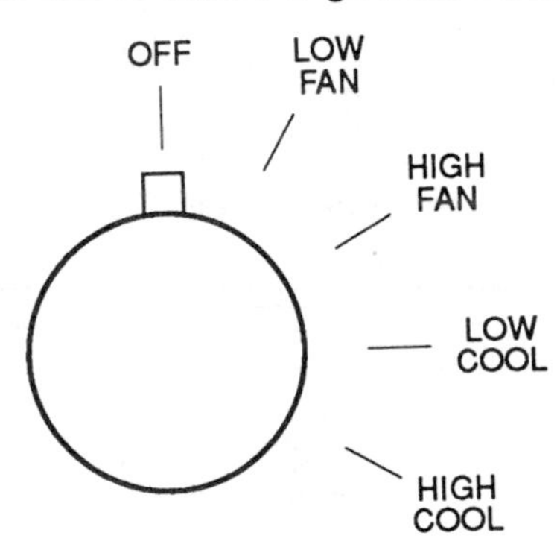

11. Find the angle through which this conduit has been bent. __________

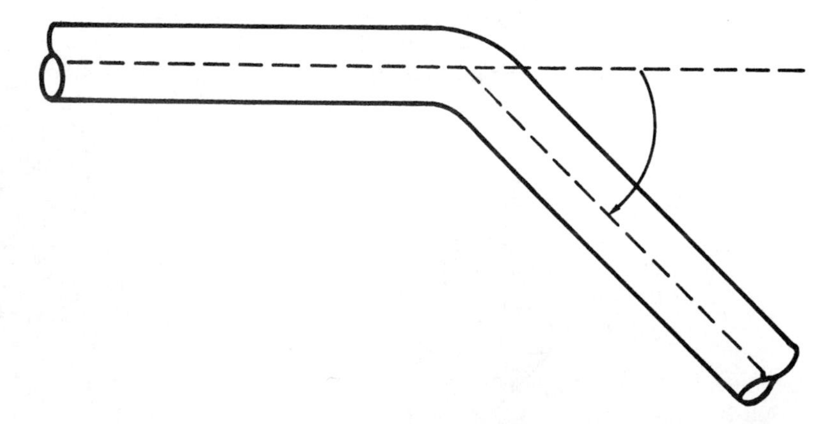

12. This damper is in the chimney flue of an oil burner. Through how many degrees is it turned from its closed position? __________

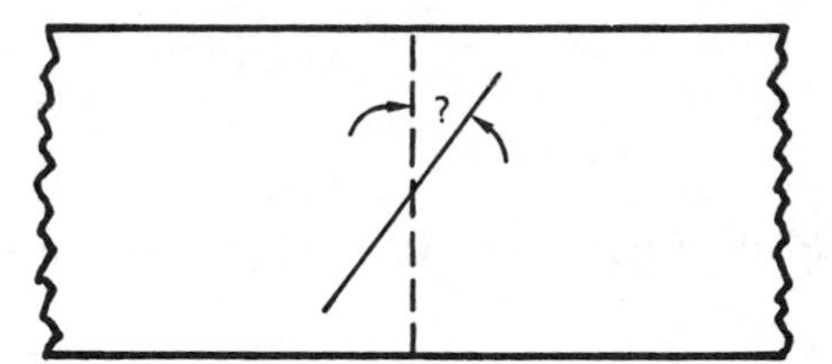

13. What is the angle between the cylinders of this compressor? __________

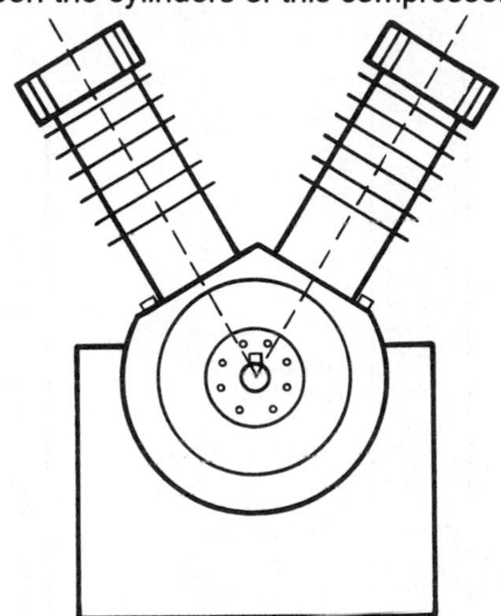

14. A bolt is turned from 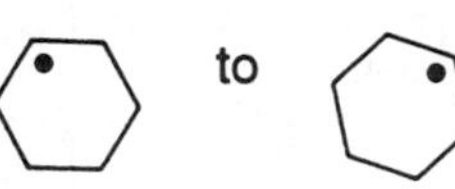 to when tightening down on it.

Through what angle was the bolt turned? __________

15. The fan of a condenser has a circular protective grill. The grill is held to the frame by 8 equally spaced bolts. How many degrees are there between the centers of the bolts? There are 360° in a circle. __________

16. When a hexagonal head bolt is turned through 60°, it looks the same as when it started. Doing that is sometimes referred to as turning the bolt 1 flat (the flat sides line up again). When tightening down on a flange to stop a leak at a cooling system heat exchanger, one bolt was turned 2 1/2 flats. Through what angle was the bolt turned? __________

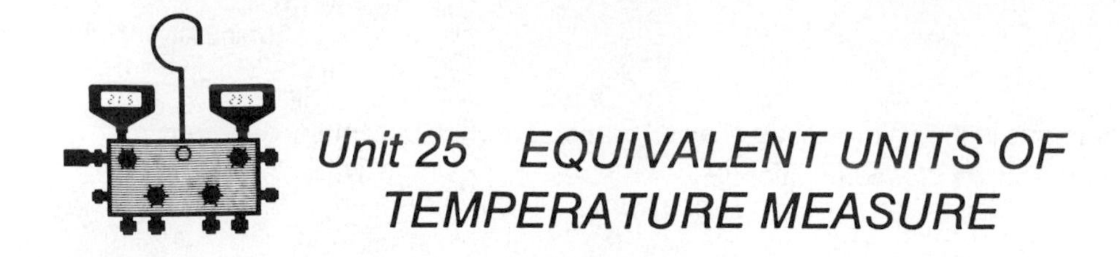

Unit 25 EQUIVALENT UNITS OF TEMPERATURE MEASURE

BASIC PRINCIPLES

- Study and apply these principles of equivalent units of temperature measure.

  Three of the most widely used temperature scales are the Fahrenheit scale, the Celsius scale, and the Kelvin scale.

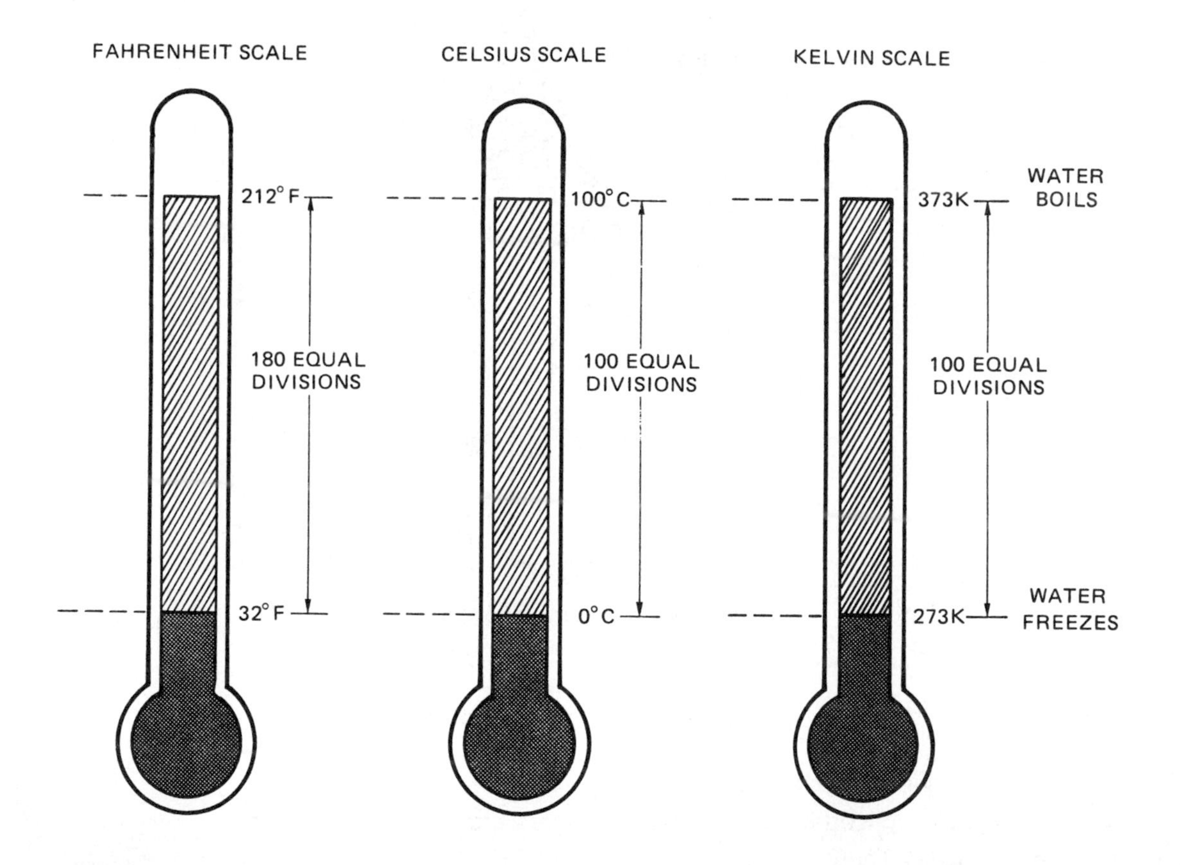

Most temperatures are measured using two different scales, the Fahrenheit and the Celsius scales. There may be times when you need to know what the temperature is on one of the scales when you have the value on the other scale.

There are two formulas to convert between the temperature scales. To convert from the Celsius scale to the Fahrenheit scale use

$$F = \frac{9}{5}C + 32$$

If the reading is in Fahrenheit degrees and you want to find the equivalent value on the Celsius scale use

$$C = \frac{5}{9}(F - 32)$$

You must be careful when using these formulas to be sure to do the correct mathematical operations in the correct order. In the first formula where there are no parentheses, make sure that the C value is multiplied by $\frac{9}{5}$ and then the 32 is added. However, in the second formula, the F value has 32 subtracted from it first and then the result is multiplied by $\frac{5}{9}$ (notice that the F – 32 is in parentheses). You must do the operations in these orders.

As most people know, temperatures can go below 0. This is often seen with outside temperatures in winter in the north or in freezers all year round. A negative (minus) sign with a temperature reading means that the temperature is less than zero.

Looking at the temperature scales, one can see when the Fahrenheit scale is at 32°, the Celsius scale is at 0°. If the temperature goes colder, the Fahrenheit number goes lower, while the Celsius value becomes a negative number. The conversion formulas give these results also.

As an example: Convert 14°F to °C.

$$°C = \frac{5}{9}(°F - 32)$$

$$= \frac{5}{9}(14 - 32) \qquad \text{Now } 14 - 32 = -18$$

$$= \frac{5}{9}(-18) \qquad \text{(The minus sign shows a bigger number was subtracted from a smaller one.)}$$

$$= -10$$

So 14°F is the same temperature as –10°C.

As another example: Convert –20°C to °F.

$$\begin{aligned} °F &= 1.8 \times °C + 32 \\ &= 1.8 \times (-20) + 32 \\ &= -36.0 + 32 \\ &= -4 \end{aligned}$$

In this case, –20°C is the same as –4°F.

It is important to be able to convert temperatures, since they need to be converted to Celsius and then to Kelvin (or absolute) in order to correctly work the gas laws that you will be seeing later in the book.

The formulas used to convert Celsius to Kelvin and Kelvin to Celsius are as follows:

$$°C = K - 273$$

$$K = °C + 273$$

Note: When converting from Fahrenheit to Celsius, be sure to subtract first, then multiply. When converting from Celsius to Fahrenheit, be sure to multiply first, then add.

PRACTICAL PROBLEMS

Express each Fahrenheit scale temperature as a Celsius scale temperature. Round to the nearer tenth when necessary.

1. 77°F ______
2. 657°F ______
3. 172°F ______
4. 127°F ______

Express each Celsius scale temperature as a Fahrenheit scale temperature.

5. 85°C ______
6. 62°C ______
7. 176°C ______
8. 228°C ______

Express each Celsius scale temperature as a Kelvin scale temperature.

9. 120°C ______
10. 35°C ______
11. 23°C ______
12. Express 41°F in kelvins. ______
13. The temperature reading on this outdoor thermometer is in degrees Fahrenheit. What is the temperature reading in degrees Celsius? ______

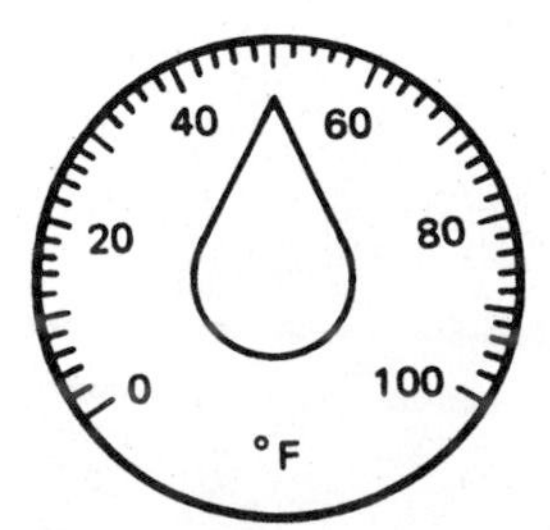

14. The water in a hot water heating system is heated to 76°C. What is this temperature reading on a Fahrenheit thermometer? __________

15. This thermostat setting is in degrees Fahrenheit. What is the equivalent thermostat setting on the Celsius scale? __________

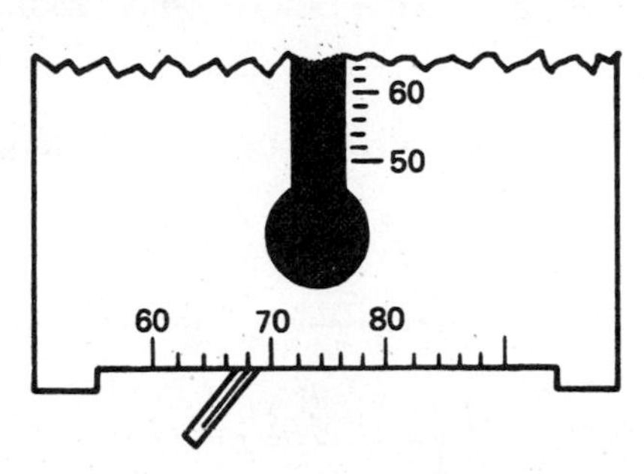

16. The temperature at which a fuel begins to burn is called the *ignition temperature*. The ignition temperature for #2 fuel oil is 700°F. Find the ignition temperature in degrees Celsius. Round the answer to the nearer tenth. __________

17. The temperature of the returned air of this warm air furnace is 65°F. The heated air is at a temperature of 140°F.

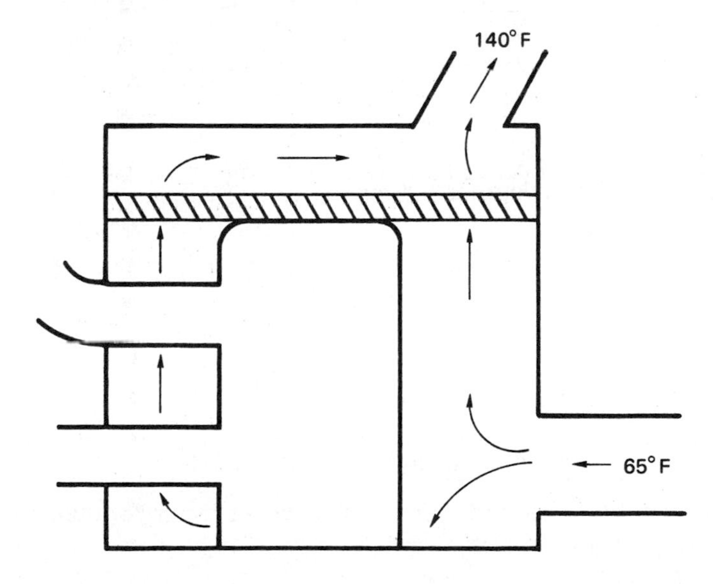

   a. What is the temperature, to the nearer tenth, of the returned air expressed in degrees Celsius? a. __________

   b. What is the temperature of the heated air expressed in degrees Celsius? b. __________

18. An air conditioner turns on when the air reaches a temperature of 80°F. The unit turns off when the air reaches 78°F.
    a. At what Celsius scale temperature, to the nearer tenth, will the unit turn on? a. ___________
    b. At what Celsius scale temperature, to the nearer tenth, will the unit turn off? b. ___________

19. The temperatures on this indoor-outdoor thermometer are in degrees Celsius.

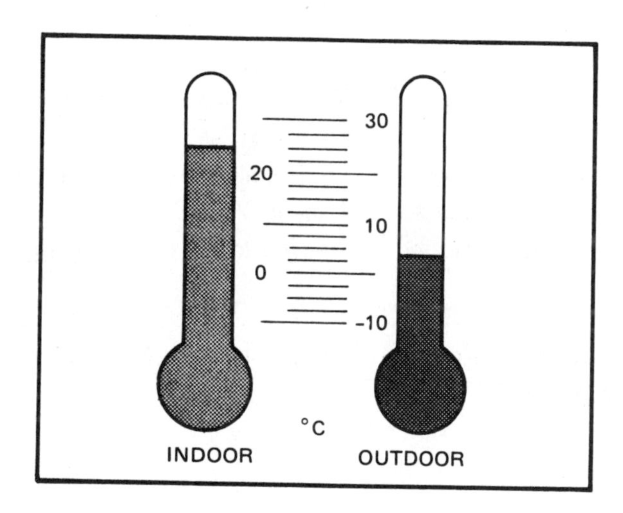

a. What is the inside temperature in degrees Fahrenheit? a. ___________

b. What is the outside temperature in degrees Fahrenheit? b. ___________

Note: Use this information for problems 20–22.

To express a temperature difference on the Fahrenheit scale as a temperature difference on the Celsius scale, use the equivalence:

$$\text{difference in °C} = \frac{5}{9} \times \text{difference in °F}$$

To express a temperature difference on the Celsius scale as a temperature difference on the Fahrenheit scale, use the equivalence:

$$\text{difference in °F} = \frac{9}{5} \times \text{difference in °C}$$

20. As a refrigerant travels through the cooling system of a refrigerator, the temperature of the refrigerant changes.

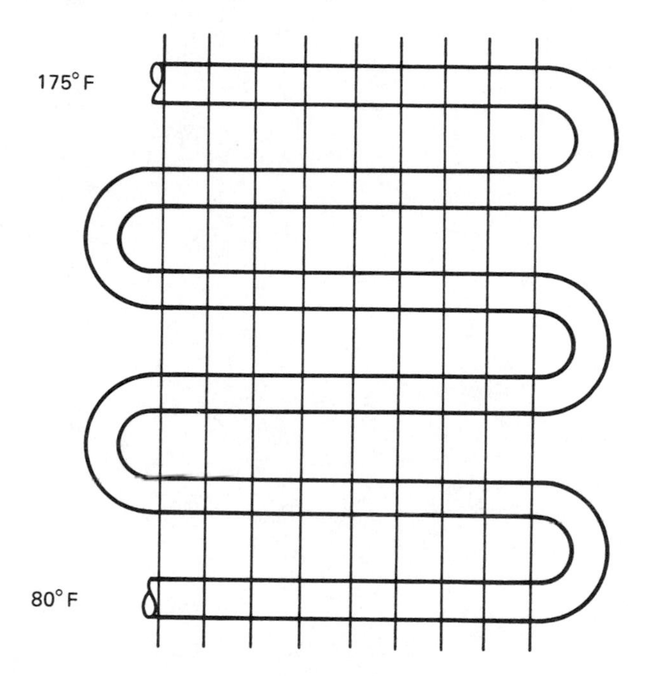

a. Find, in degrees Fahrenheit, the temperature difference. a. ____________

b. Find, in degrees Celsius, the temperature difference. Round the answer to the nearer tenth. b. ____________

21. The temperature difference between the floor and the ceiling of a room is 8°F. Express this difference in degrees Celsius. Round the answer to the nearer tenth. ____________

22. On a cold day, the temperature difference between the inside and the outside of a certain house is 25°C. Express this value on the Fahrenheit scale. ____________

23. When R-21 expands in the refrigeration cycle, its temperature drops to 5°F. What is this temperature on the Celsius scale? ____________

24. During the refrigeration cycle, R-12 is compressed. Its temperature reaches 170°F. What is the Celsius equivalent to this temperature? ____________

25. A freezer is to be maintained at 20°F. What would a Celsius thermometer read when placed in this freezer? ____________

Computed Measure

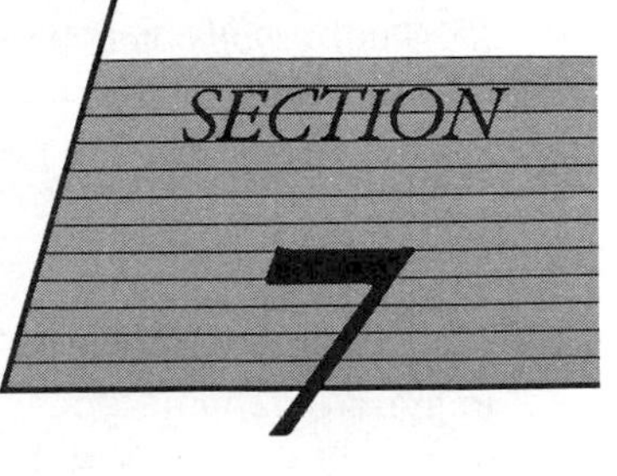

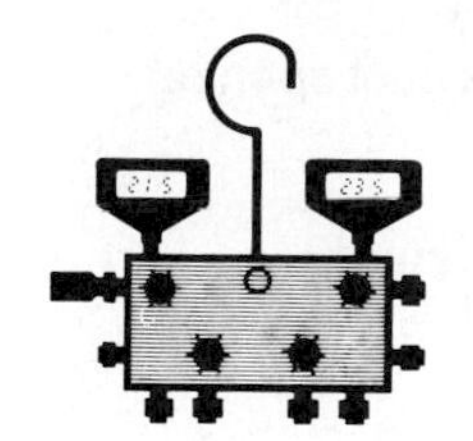

Unit 26 AREA MEASURE

BASIC PRINCIPLES

- Review and apply the principles of area measure to the problems in this unit.
- Review denominate numbers in section I of the appendix.
- Use formulas for areas located in section III of the appendix.

The amount of space on the surface of a figure is called the area. Area is also the number of square units equal in measure to the surface of a figure. Many figures have regular shapes to them. Formulas have been developed to find the area of these regular figures. The various formulas will be given throughout the unit.

To solve for area, determine the correct formula to use and then substitute values in for the terms in the formula. Be certain that the units are the same. The units of area are square inches, square feet, square meters, or similar units.

Note: When solving any of these problems, make a little table of each dimension and its measurement. Then write the correct formula, substitute into it as the next step, and then solve it. Be sure the units are all the same.

PRACTICAL PROBLEMS

Note: Use this information for problems 1–3. The area of a square is:

area = side × side

or

$A = s \times s$

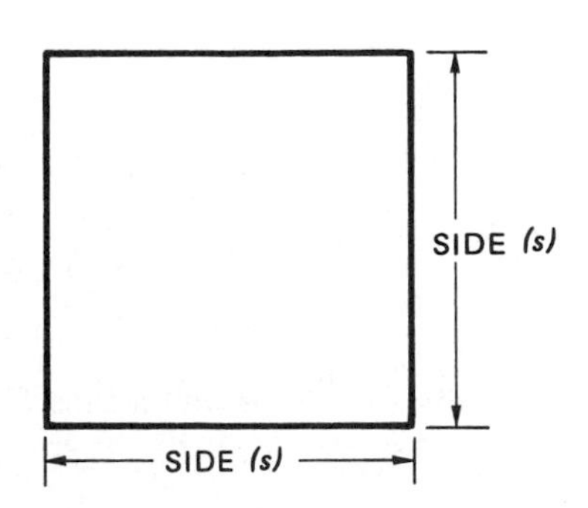

1. The sides of a square duct are 8 inches. Find, in square inches, the area of the duct opening.

2. A square duct has sides of 24 centimeters. What is the area of the duct opening in square centimeters.

3. This grill at the end of an air duct has equally spaced square openings.

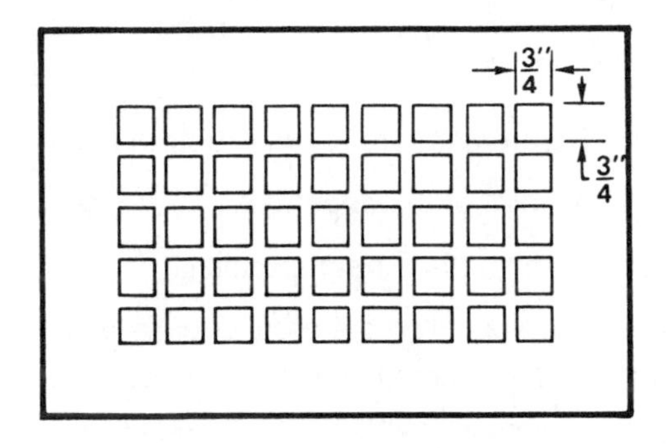

   a. Find, in square inches, the area of each opening. a. ____________
   b. What is the total area of the openings through which air can flow? b. ____________

Note: Use this information for problems 4–9.
The area of a rectangle is:

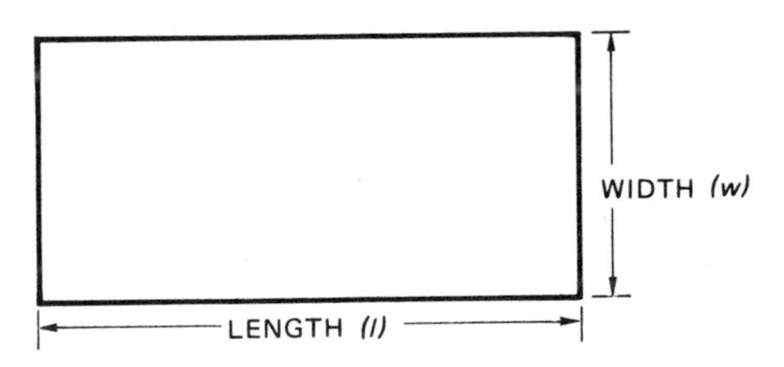

area = length × width

or

$A = l \times w$

4. Find in square meters, the area of a wall that is 5 meters long and 3 meters high. ____________

5. The indicated wall must be painted after insulation is installed. Find, in square feet, the area of this wall so the proper amount of paint can be ordered. ______

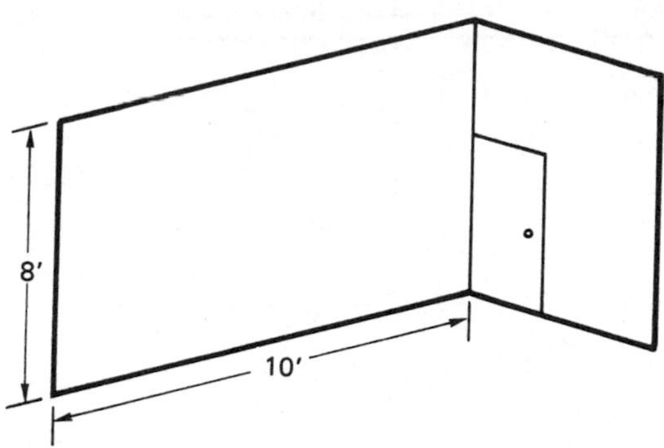

6. What is the area of this window? Express the answer in square meters. ______

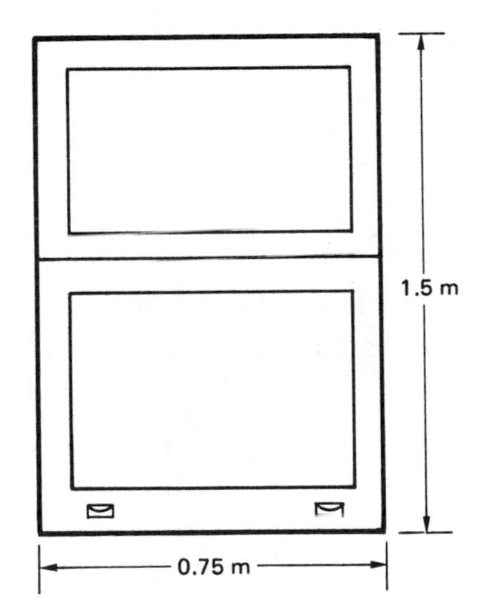

7. Cold air returns to a forced air heating system through 3 air ducts. Each of the ducts is rectangular and measures 16 inches by 20 inches. What is the total area of the return air duct openings? ______

8. This house plan shows the dimensions of each room. Find, in square meters, the floor area of each room.

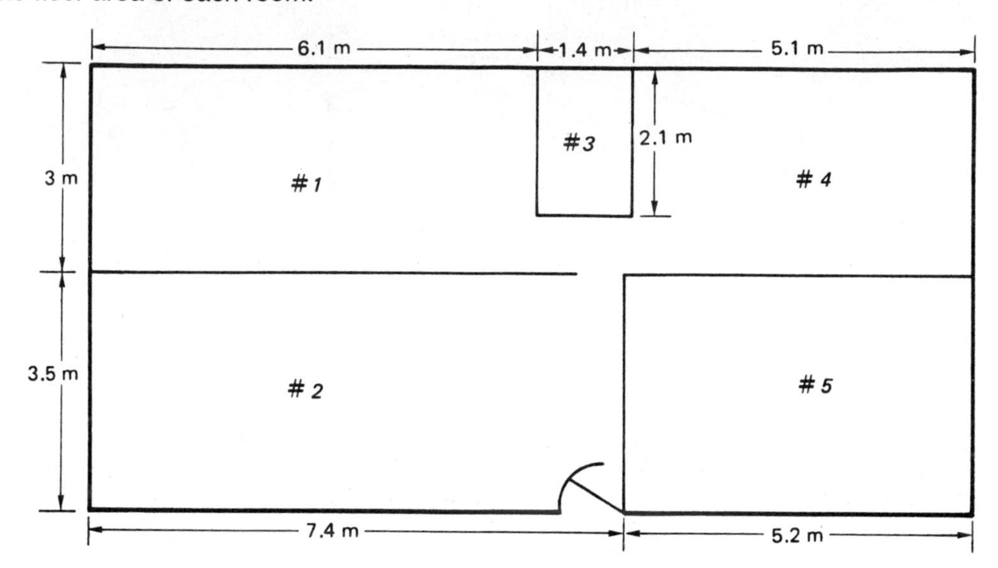

a. Room #1 a. ____________
b. Room #2 b. ____________
c. Room #3 c. ____________
d. Room #4 d. ____________
e. Room #5 e. ____________
f. What is the total area of the house? f. ____________

9. Styrofoam sheeting is used as insulation in the walls of refrigerators. Each wall is 0.55 meter by 1.6 meters. How many square meters of sheeting are needed for seven walls? ____________

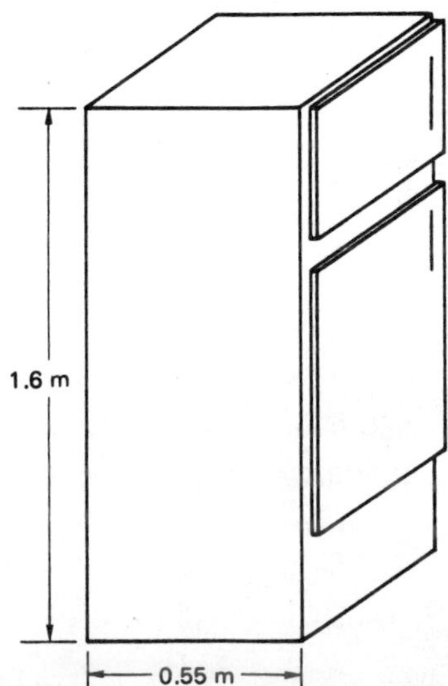

Note: Use this information for problems 10–13.

The area of a circle is:

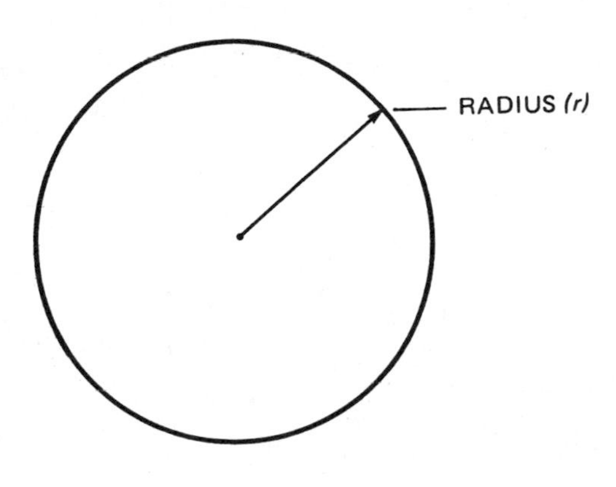

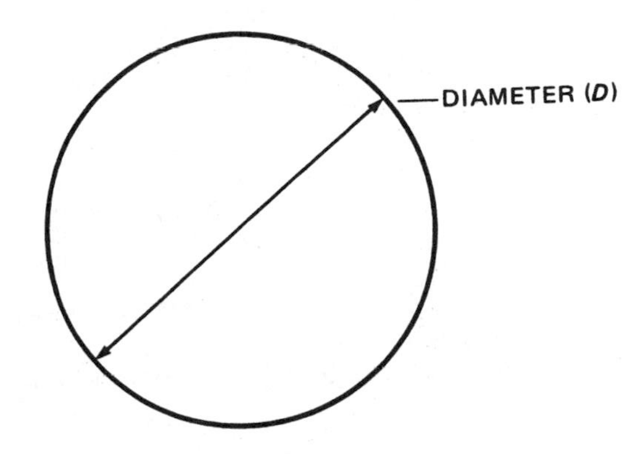

$A = \pi \times r \times r$ *or* $A = \frac{\pi}{4} \times D \times D$

where $\pi = 3.1416$

This means that the area of a circle is:

$A = 3.1416 \times r \times r$ *or* $A = \frac{3.1416}{4} \times D \times D$

or

$A = 0.785\,4 \times D \times D$

10. Find, in square centimeters, the area of the opening in this circular duct. Round the answer to the nearer thousandth. __________

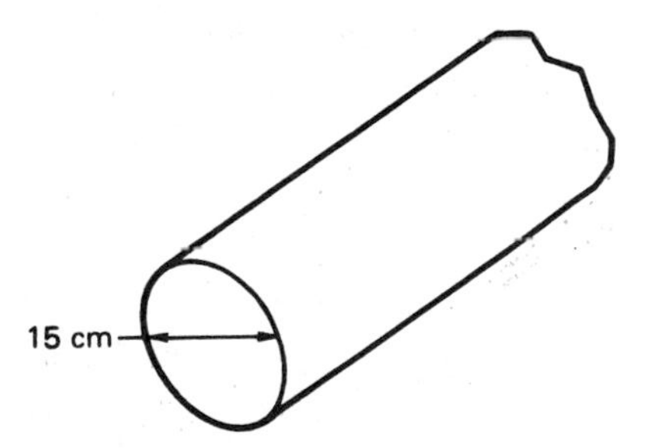

11. What is the area of the opening in a duct which has a diameter of 7 inches? Round the answer to the nearer thousandth square inch. __________

12. The cylinder of a compressor has a diameter of 2.4 centimeters. What is the area of the opening in the cylinder? Round the answer to the nearer hundredth square centimeters. __________

13. Find, to the nearer hundredth square inch, the area of the top of this piston for a compressor. ______

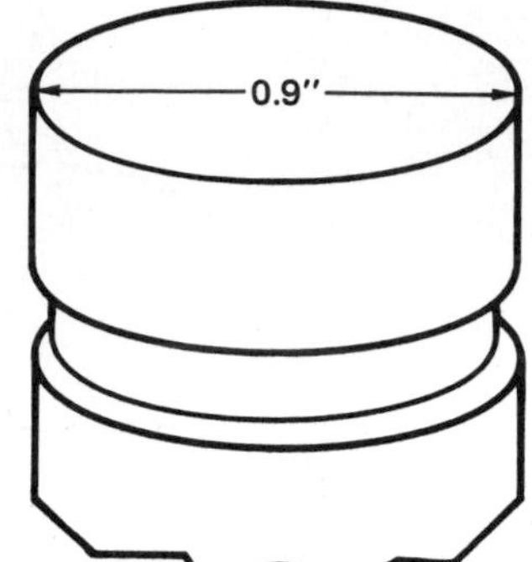

Note: Use the information for problems 14 and 15.

The area of a triangle is:

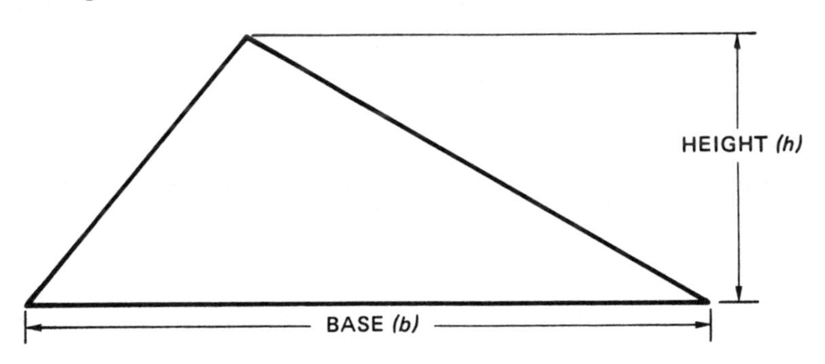

$$\text{area} = \frac{1}{2} \times \text{base} \times \text{height}$$

or

$$A = \frac{1}{2} \times b \times h$$

14. How many square centimeters of metal are needed to make this shield? ______

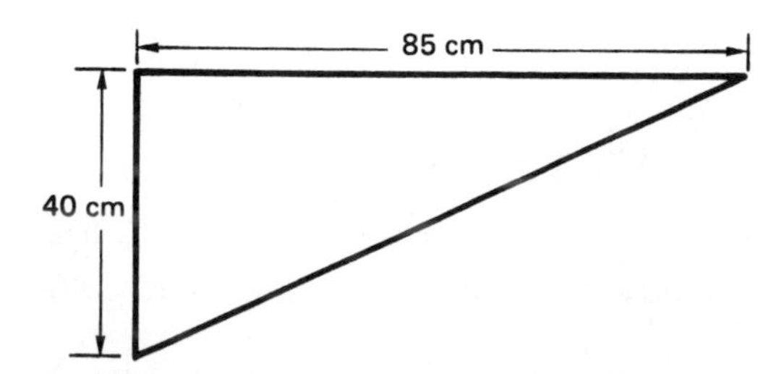

15. The bedroom of an A-frame house has a triangular-shaped wall. What is the area of the wall including the door? ____________

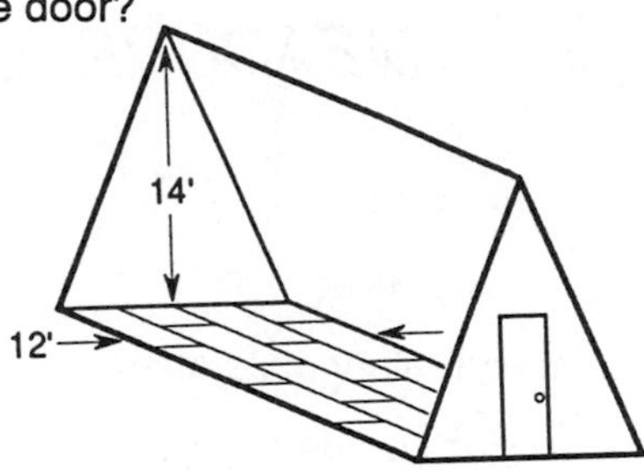

16. What is the area of this floor in square meters? ____________

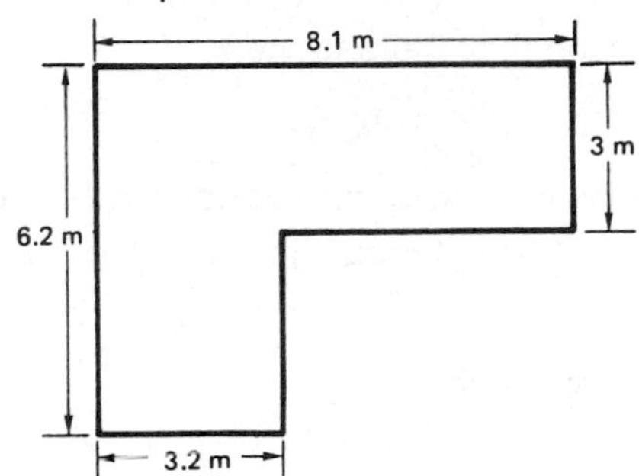

17. A rectangular duct must connect with a circular duct. The rectangular duct measures 6 inches by 7 inches. The circular duct has a diameter of 7 inches. Which duct has the smaller cross-sectional area? ____________

Note: Use this information for problems 18 and 19.

The area of a rectangle is $A = l \times w$.
This means that the length of the rectangle is the area (A) divided by the width (w) or $l = \frac{A}{w}$.

18. For proper airflow, a rectangular duct must have an area of 123.25 square centimeters. The duct must be placed in the wall and must have a width of 8.5 centimeters. What must the length of the duct opening be? ____________

19. The area of a duct opening must be 84 square inches. The duct must be placed above a ceiling and must be 7 inches high. What length must the duct opening be? ____________

20. An 8-inch by 12-inch rectangular duct splits into two branch ducts. The area of the two branches is equal to the area of the 8-inch by 12-inch duct. One of the branches is a square duct measuring 6 inches on each side. What is the area of the opening in the second branch? ____________

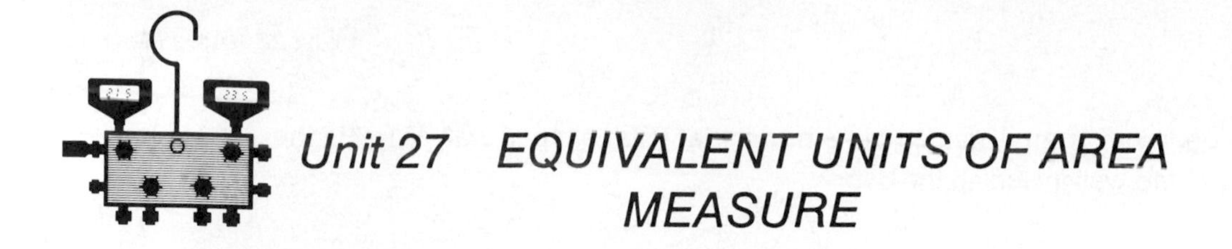

Unit 27 EQUIVALENT UNITS OF AREA MEASURE

BASIC PRINCIPLES

- Review and apply the principles of equivalent units of area measure to the problems in this unit.
- Review denominate numbers in section I of the appendix.
- Study tables of equivalent units of area measure.
- Review formulas for area in section III of the appendix.

Converting from one set of units of area measure to another is done in the same manner as linear measure. Fractions are formed to remove the current unit of measure and bring in a new unit. The only differences are the equivalent values.

The new equivalent values are in the table below.

ENGLISH AREA MEASURE		
1 square yard (sq yd)	=	9 square feet (sq ft)
1 square foot (sq ft)	=	144 square inches (sq in)

METRIC AREA MEASURE		
100 square millimeters (mm^2)	=	1 square centimeter (cm^2)
10 000 square centimeters (cm^2)	=	1 square meter (m^2)
1 000 000 square meters (m^2)	=	1 square kilometer (km^2)

Note: When solving any of these problems, make a little table of each dimension and its measurement. Then write the correct formula, substitute into it as the next step, and then solve it. Be sure the units are all the same.

PRACTICAL PROBLEMS

1. To find the heat loss of a room, the area of this window must be determined. Find, in square feet, the area of this window. ____________

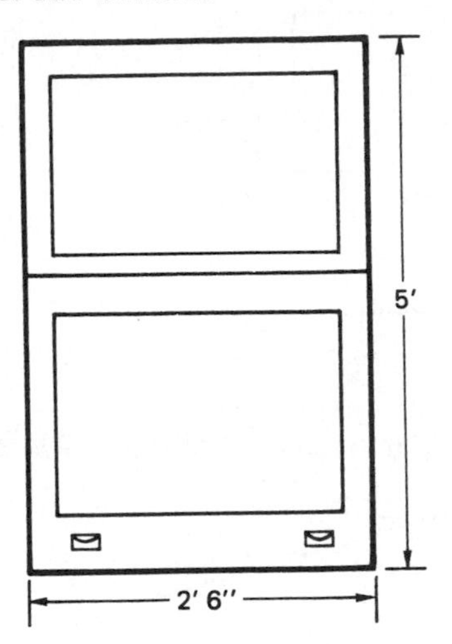

2. A roll of 3 1/2-inch thick insulation contains 30 feet of insulation. The insulation is 16 inches wide.
   a. Find the number of square inches of space the insulation will cover. a. ____________
   b. Find the number of square feet the insulation will cover. b. ____________

3. A rectangular air duct measures 15 centimeters by 25 centimeters.
   a. What is the area of the opening in the duct in square centimeters. a. ____________
   b. What is the area of the opening in the duct in square meters? b. ____________

4. While repairing a baseboard hot water heating system, a repairer drops a lighted blow torch, badly damaging the rug. How many square yards of carpeting will be needed to replace the damaged rug? Round off to the next higher full square yard. ____________

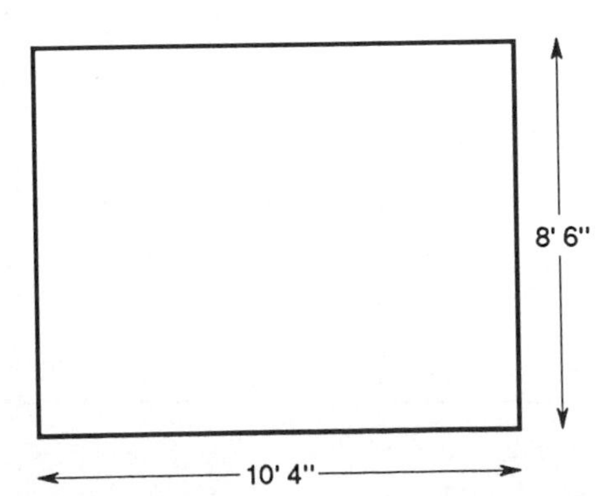

5. This wall is to be insulated. Find the number of square feet of insulation that is needed. __________

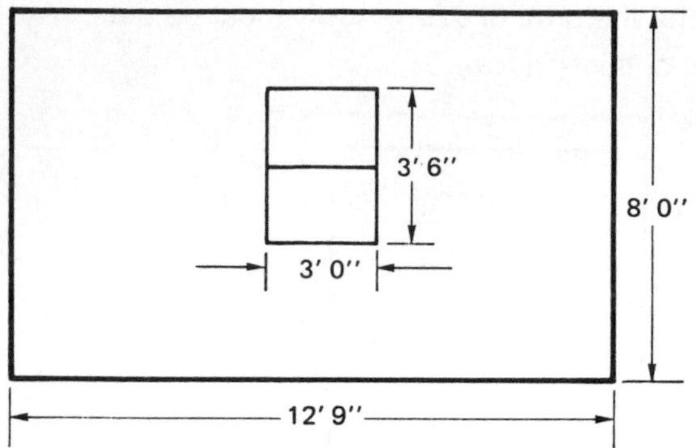

6. The ceiling in a room is sloped. What is the area of the wall in square feet? __________

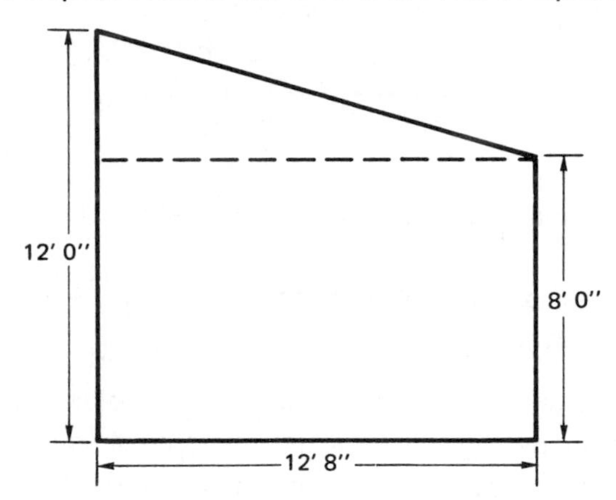

7. To reduce the heat load in a house during the summer, a reflective coating is applied to the 2 windows shown below. How many square inches of glass must be coated? __________

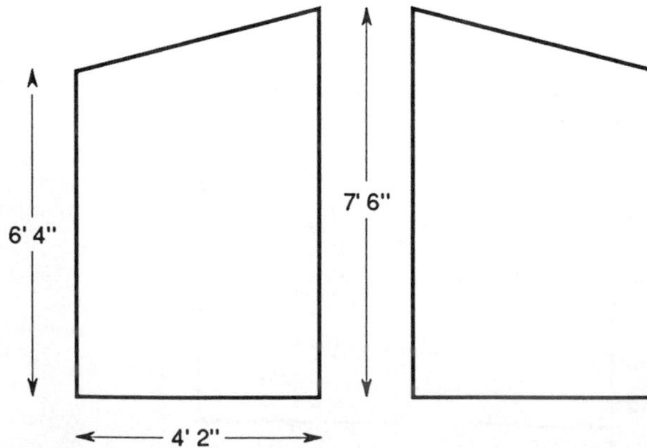

8. One of the 250 fins of a baseboard heater is shown. Each fin has a 2.5-centimeter hole in it.

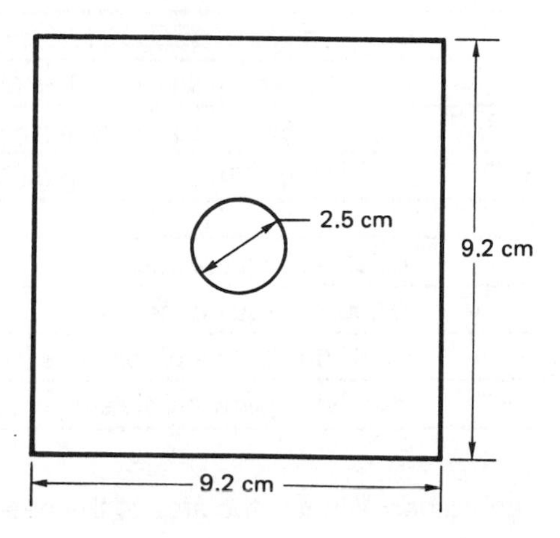

a. Find, in square centimeters, the area of the 250 fins. a. ____________

b. Find, in square meters, the area of the 250 fins. b. ____________

9. This filter is placed in a forced air system. What is the area through which the air can flow? Round the answer to the nearer hundredth square inch. ____________

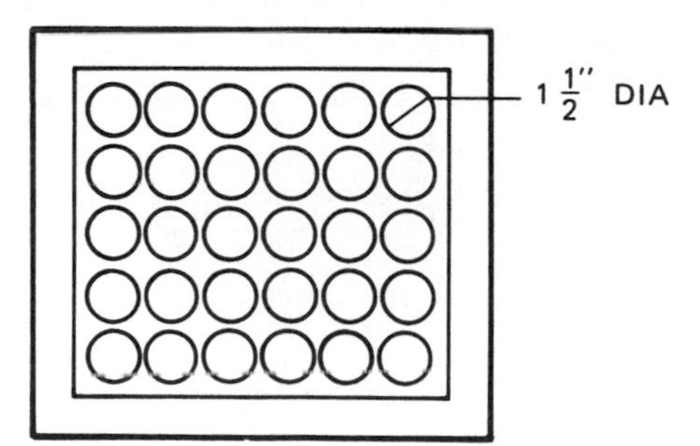

10. What is the area of the opening of a 7-inch round duct to the nearer tenth square foot? ____________

Note: Use these tables of equivalent units of area measure for problems 11–15.

1 square meter (m^2)	=	10.763 910 square feet (sq ft)
1 square meter (m^2)	=	1 550.000 square inches (sq in)
1 square centimeter (cm^2)	=	0.155 000 square inch (sq in)
1 square millimeter (mm^2)	=	0.001 550 square inch (sq in)

1 square foot (sq ft)	=	0.092 903 square meter (m^2)
1 square inch (sq in)	=	0.000 645 square meter (m^2)
1 square inch (sq in)	=	6.451 600 square centimeters (cm^2)
1 square inch (sq in)	=	645.160 square millimeters (mm^2)

11. The opening in an air duct is 72 square inches. What is the area to the nearer hundredth square centimeter? __________

12. A wall has an area of 75 square feet. Find, to the nearer hundredth square meter, the area of the wall. __________

13. The filter for a room air conditioning unit has an area of 1 600 square centimeters. How many square inches are there in the filter? __________

14. The top of a piston for a compressor has an area of 3.4 square inches. Find the area to the nearer thousandth square centimeter. __________

15. The installation instructions for an imported condensing unit for a domestic heat pump system states that it should sit on a slab at least 1.2 square meters in area. What is the minimum size of the slab in square feet? Round off to the nearer tenth square foot. __________

Unit 28 RECTANGULAR VOLUMES

BASIC PRINCIPLES

- Review and apply the principles of volume of rectangular solids to the problems in this unit.
- Review denominate numbers in section I of the appendix.
- Review the tables of equivalent units of length measure in section II of the appendix.
- Study tables of equivalent units of volume measure.
- Use formulas for volumes located in this unit and in section III of the appendix.
- Review the tables of equivalent units of volume measure in section II of the appendix.

Volume is the space enclosed by a three-dimensional figure. Volumes are found by multiplying the length of a figure by its width and then multiplying that number by the figure's height. It can sometimes be found by multiplying the area of one surface by the depth of the figure from that surface. The units for volume are cubic units, such as cubic inches, cubic feet, or cubic meters.

A rectangular solid is a box-like figure. The volume of a rectangular solid is:

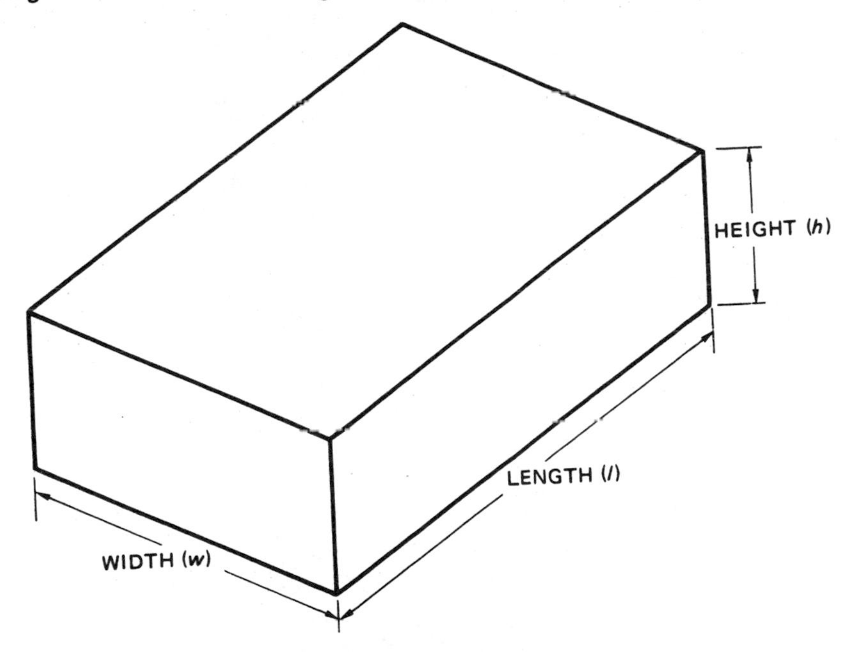

volume = length × width × height

or

$$V = l \times w \times h$$

Again, care must be taken to make sure that all of the units are the same.

If units do have to be changed, use equivalent units given in the tables below.

ENGLISH VOLUME MEASURE		
1 cubic yard (cu yd)	=	27 cubic feet (cu ft)
1 cubic foot (cu ft)	=	1 728 cu inches (cu in)

METRIC VOLUME MEASURE		
1 000 cubic millimeters (mm^3)	=	1 cubic centimeter (cm^3)
1 000 000 cubic centimeters (cm^3)	=	1 cubic meter (m^3)

Note: When solving any of these problems, make a little table of each dimension and its measurement. Then write the correct formula, substitute into it as the next step, and then solve it. Be sure the units are all the same.

PRACTICAL PROBLEMS

1. Find, in cubic feet, the volume of this room. __________

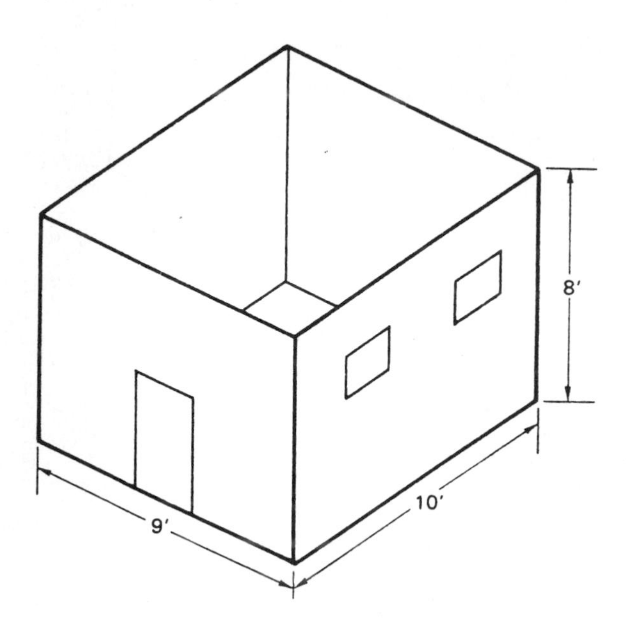

2. The condenser for a heat pump system sits on this slab of concrete. How many cubic centimeters of concrete are needed to make this slab? ____________

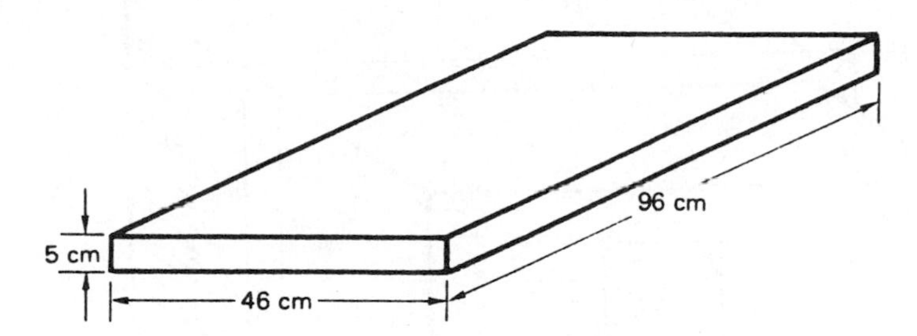

3. Find, in cubic meters, the volume of this horizontal oil furnace. ____________

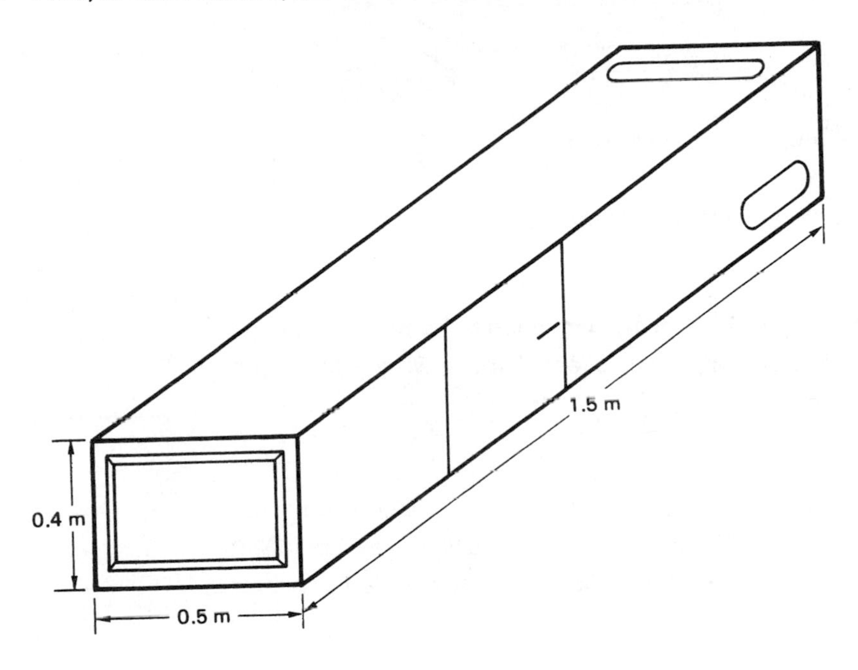

4. A mobile home measures 14 feet wide and 70 feet long. It has 7-foot high ceilings. What is the approximate volume to be air conditioned for this mobile home? ____________

5. The dimensions of this window air conditioning unit are in inches.

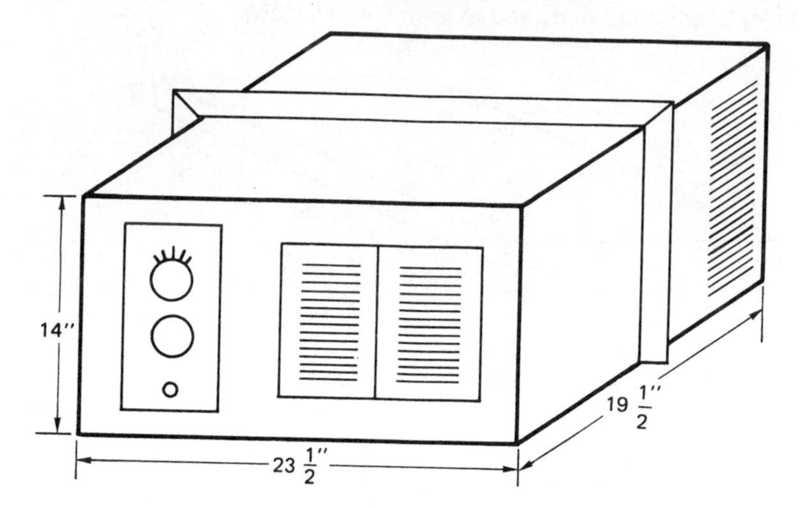

a. Find, in cubic inches, the volume of the unit. a. ____________

b. What is the volume of the unit to the nearer hundredth cubic foot? b. ____________

6. R-19 insulation is 6 inches thick. A roll of this insulation is 15 inches wide and has 32 feet of insulation in it. How many cubic feet of space will this insulation fill? ____________

7. The inside dimensions of a refrigerated tractor trailer are 91 inches per side, 100 7/8 inches high, and 44 feet 1/2 inches long. Find the volume in cubic feet that must be cooled by the refrigeration unit. Round the answer to the nearer tenth cubic foot. ____________

8. A room measures 12 1/2 feet wide and 15 1/2 feet long. The walls are 8 feet high. The volume of air in the room changes six times each hour. How many cubic feet of air enters the room each minute? ____________

9. The refrigerating compartment of a refrigerator has a volume of 11.55 cubic feet. The inside of the compartment is 2.2 feet wide and 3 feet high. Using the formula $l = \frac{V}{w \times h}$, find the inside length of the refrigerating compartment. ____________

10. The walls in this room are 8 feet high.

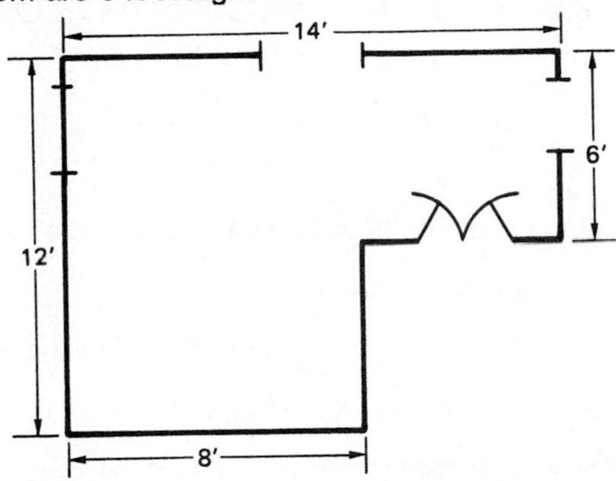

a. What is the volume of this room in cubic feet? a. ____________

b. There is 0.028 317 cubic meter of space in 1 cubic foot of space. How many cubic meters of space are in this room? Round the answer to the nearer thousandth cubic meter. b. ____________

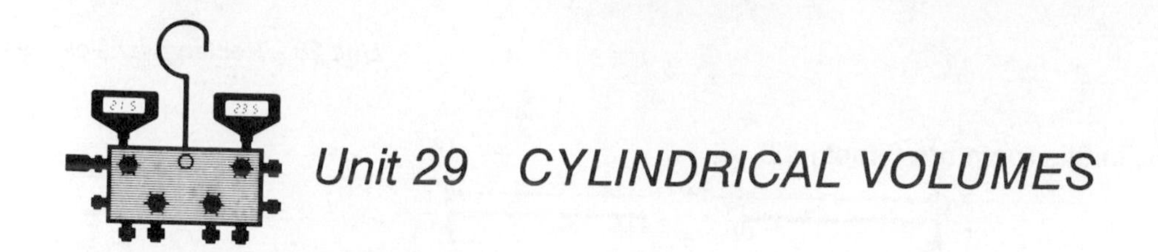

Unit 29 CYLINDRICAL VOLUMES

BASIC PRINCIPLES

- Review and apply the principles of volume of cylindrical solids to these problems.
- Review denominate numbers in section I of the appendix.
- Review the tables of equivalent units of length measure in section II of the appendix.
- Review the tables of equivalent units of volume measure in section II of the appendix.

Finding the volume of a cylindrical solid is similar to finding the volume of any solid, except a different formula is used. Cylindrical solids are solids with circular ends. The formula for finding the volume includes the area of a circle in it.

The volume of a cylindrical solid is:

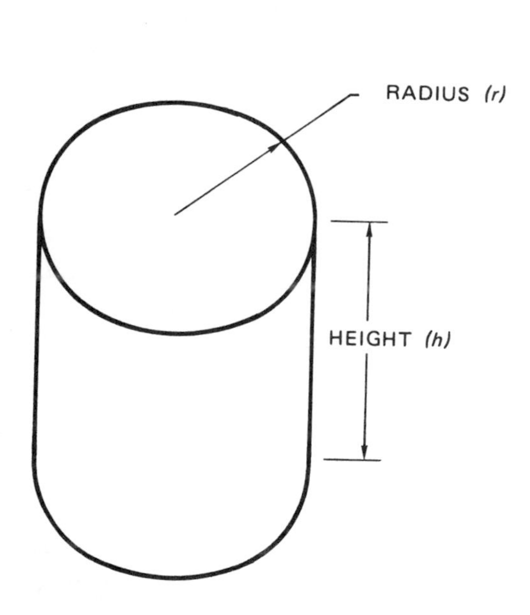

DIAMETER (D)

HEIGHT (h)

$V = \pi \times r \times r \times h$ *or* $V = \frac{\pi}{4} \times D \times D \times h$

where $\pi = 3.141\ 6$.

This means that the volume of a cylindrical solid is:

$$V = 3.141\,6 \times r \times r \times h \quad \textit{or} \quad V = \frac{3.1416}{4} \times D \times D \times h$$

or

$$V = 0.7854 \times D \times D \times h$$

Note: When solving any of these problems, make a little table of each dimension and its measurement. Then write the correct formula, substitute into it as the next step, and then solve it. Be sure the units are all the same.

PRACTICAL PROBLEMS

Note: Round all answers to three decimal places.

1. A cylindrical tank for refrigerant has an inside diameter of 14 inches and is 16 inches high. What is the volume of the tank in cubic inches? __________

2. Find, in cubic centimeters, the volume of this compressor. __________

3. A cylinder containing propane has an inside diameter of 2.5 inches and is 10 inches long. How many cubic inches of propane can the container hold? __________

4. The combustion chamber of an oil furnace is a cylinder. The chamber is 18 inches long and has an inside diameter of 14 inches.

   a. What is the volume of the chamber in cubic inches? a. __________

   b. What is the volume of the chamber in cubic feet? b. __________

5. The hot water system in a house uses pipes which have an inside diameter (I.D.) of 1 inch. There are 25 feet of pipes in one part of the system. How many cubic feet of water can these pipes hold? ________

6. This fuel tank is 60 inches long.

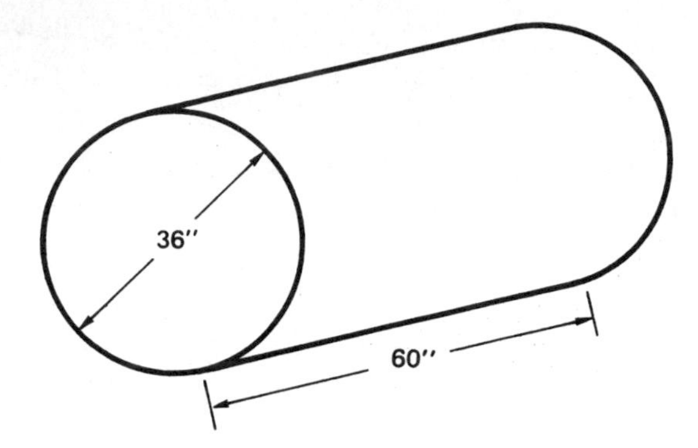

   a. Find, in cubic inches, the volume of the tank. a. ________
   b. One gallon of fuel will occupy 231 cubic inches of space. How many gallons of fuel will this tank hold? b. ________

7. The inside dimensions of a rectangular oil furnace are 60 centimeters wide by 1.15 meters long by 45 centimeters high. The combustion chamber is cylindrical, is 46 centimeters long, and has an outside diameter of 43 centimeters. How many cubic centimeters of space are there between the inside walls of the burner and the outside walls of the combustion chamber? ________

8. The inside diameter of the cylinder of the compressor for a heat pump is 5 centimeters. The cylinder is 3.65 centimeters long.
   a. Find, in cubic centimeters, the volume of the cylinder. a. ________
   b. There is 0.061 024 cubic inch of space in one cubic centimeter of space. How many cubic inches of space are there in this compressor cylinder? Round the answer to the nearer tenth. b. ________

9. A load of 50 rolls of R-11 insulation must be picked up. Each roll is 25 inches in diameter and 15 inches tall. What is the smallest volume a closed-back truck could hold to be able to pick these rolls up in one load? (In reality, a larger volume is needed since the truck is rectangular and the rolls are cylindrical.) Round off the answer to the next higher whole cubic foot. ________

10. An old cylindrical fuel oil tank is being removed from a work site. It is 70 inches long and 40 inches in diameter. It is half full of Grade 2 fuel oil. Each cubic foot of fuel oil weighs 53.1 pounds. If the tank itself weighs 215 pounds, what is the total weight to be carried by the truck removing the tank? Round off the answer to the nearer tenth pound. __________

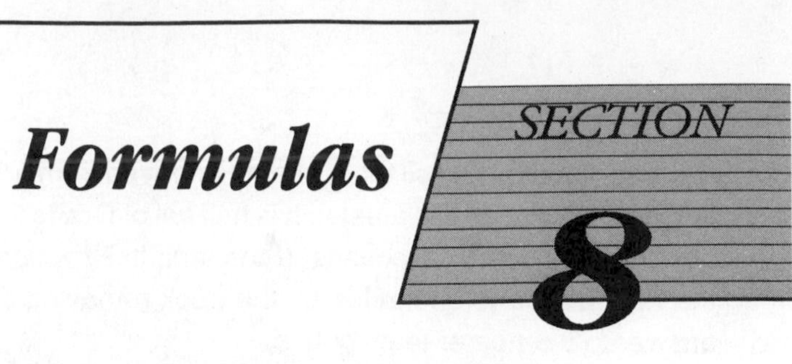

Formulas

Unit 30 OHM'S LAW AND ELECTRICAL RELATIONSHIPS

BASIC PRINCIPLES

- Study and apply these principles of Ohm's Law to the problems in this unit.

Formulas are relationships that can be used over and over again by changing the values that are put into them. In formulas, letters represent numbers. The letters represent the same quantity—not the same value—all of the time. To solve the formula, you substitute numbers in for those letters. Care must be taken to substitute the correct value in for the correct letter. Care must also be taken to have the correct units with the values that are being substituted.

You have already been working with formulas. In the last several units, formulas were used to find areas and volumes. There are many different formulas that you will have to work with in the heating and air conditioning field. In this unit formulas dealing with electrical values will be studied.

Current is produced by electrons traveling from one point to another. It is the flow of electric charge. The unit measure for electrical current is *amperes*.

The flow of electrons is dependent upon the voltage of the system and the resistance. *Resistance* is the opposition to the flow of electric charge. It is measured in *ohms* (Ω). The *voltage* is the force applied to cause the electrons to flow through the resistance. It is measured in *volts*.

The relationship between the current (I), voltage (E), and the resistance (R) is known as *Ohm's Law*.

$$\text{current } (I) = \frac{\text{voltage } (E)}{\text{resistance } (R)}$$

The law states that the current (I) is directly proportional to the force (E) applied to produce the current and indirectly, or inversely, proportional to the resistance (R) to the flow of electrons. This means that the higher the voltage, the higher the current; and the smaller the resistance, the higher the current.

Ohm's Law may also be written as:

$$R = \frac{E}{I} \qquad \text{or} \qquad E = I \times R$$

The units for this formula are unusual. Volts = amperes (amps) × ohms.

Note: Always use the formula that has what you are looking for by itself on the left side of the equal sign.

PRACTICAL PROBLEMS

Round the answer to the nearer hundredth when necessary.

Note: Use this diagram for problems 1–6.

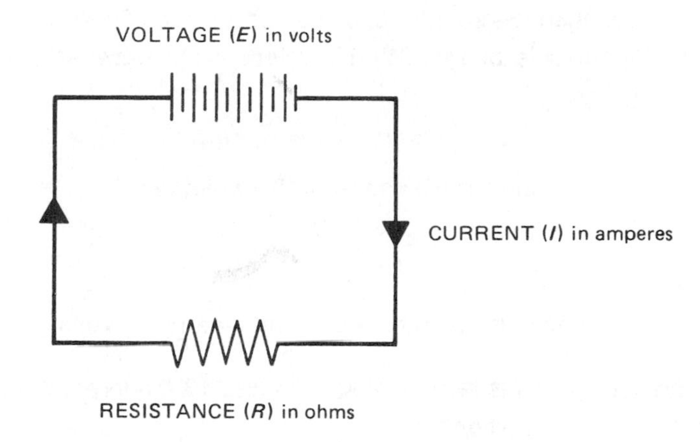

Complete this chart.

	VOLTAGE (*E*) in volts	CURRENT (*I*) in amperes	RESISTANCE (*R*) in ohms
1.	50	25	
2.		15	4
3.	25		10
4.	120	1/2	
5.	200		75
6.		5	72.5

7. A portable heater is listed for 115 volts. It uses 12 amperes of current. What is the resistance of the heater? __________

8. An electric baseboard heater has 25 ohms of resistance. The heater uses 9.6 amperes of current. Find, in volts, the voltage required for the heater. __________

9. The on-off switch light for a car air conditioner runs off the car's 12-volt battery. The light has a resistance of 16 ohms. What current passes through the bulb? ______

10. The heater of the automatic defroster in a refrigerator is a resistance wire. The heater uses 110 volts and draws 4.8 amperes of current. Find, in ohms, the resistance. ______

Note: Use this information for problems 11–14.
Power is the rate of doing work. In electricity, it is the work done when one ampere of current is pushed through a circuit by a pressure of one volt. The unit of measure for power is the *watt* (W). The relationships between power, current and voltage are:

$$\text{power } (P) = \text{current } (I) \times \text{voltage } (E)$$

or

$$I = \frac{P}{E} \qquad E = \frac{P}{I}$$

The units used here are watts = amps × volts.

11. A window air conditioning unit is rated at 230 volts and 13 amperes. What is the power used by the air conditioner? ______

12. The motor for the pump of a high pressure fuel oil gun is rated at 120 volts. The motor uses 624 watts of power. Find, in amperes, the current that the motor draws. ______

13. A furnace for an electric heating system is rated at 121 amperes and 27 500 watts. What is the voltage of this system? ______

14. A central air conditioning unit is rated at 4 100 watts. The unit uses 230 volts. How many amperes of current does the electrical cable for this unit carry when current is flowing? ______

Note: Use this information for problems 15–18.
An *electric circuit* is a conducting path that leads from the source, to energy devices or resistances, and back to the source. The current that flows through the circuit can travel in different paths.
When the electrical charge, or current, is provided with only one possible route to follow, the current is a *series circuit*. This means that the total current flows from the source, through each device, and back to the source.

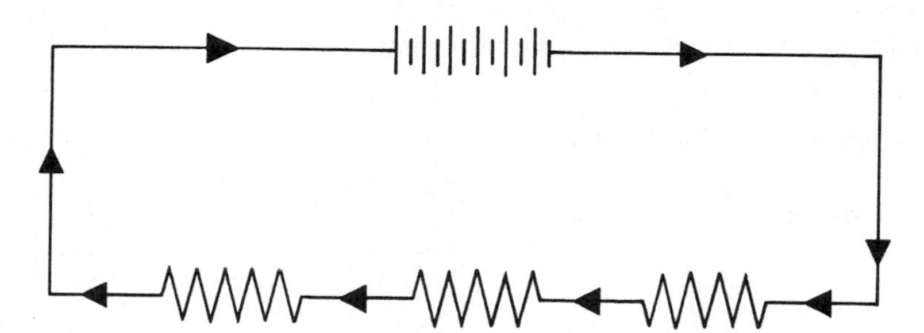

In a series circuit, the total resistance is equal to the sum of each device's resistance.

$$R_T = R_1 + R_2 + R_3 + \ldots$$

When the total current splits up and flows to the devices in separate paths, the circuit is a *parallel circuit*.

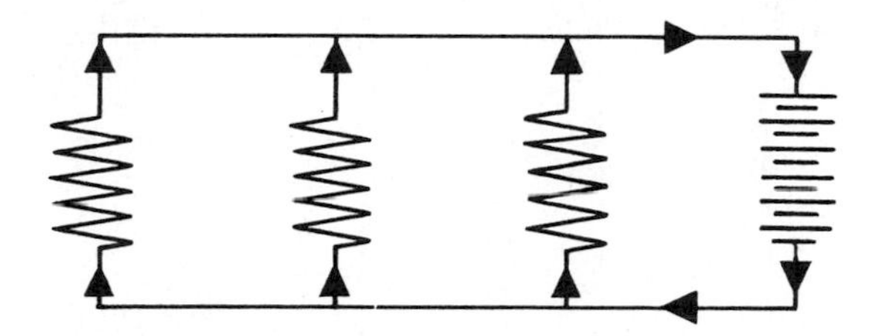

In a parallel circuit, the equivalent resistance is:

$$R_T = \frac{1}{\frac{1}{R_1} + \frac{1}{R_2} + \frac{1}{R_3} + \ldots}$$

In a series circuit, the total voltage is $E_T = E_1 + E_2 + E_3 + \ldots$

But, in a parallel circuit, $E = E_1 = E_2 = E_3 = \ldots$

In a series circuit, $I = I_1 = I_2 = I_3 = \ldots$

But, in a parallel circuit, $I_T = I_1 + I_2 + I_3 + \ldots$

Note: It is very important to realize that

$R_T = \frac{1}{\frac{1}{R_1} + \frac{1}{R_2} + \frac{1}{R_3} + \ldots}$ is a lot different from $R_T = R_1 + R_2 + R_3 + \ldots$

To use the first formula, always solve it in the following way:

1) Find each value for $\frac{1}{R_1}$, $\frac{1}{R_2}$, and so on.

2) Add those values together.

3) Then divide that total into 1.

15. Three leads of a compressor are connected as shown. What is the total resistance between **S** and **R** when **C** is disconnected? __________

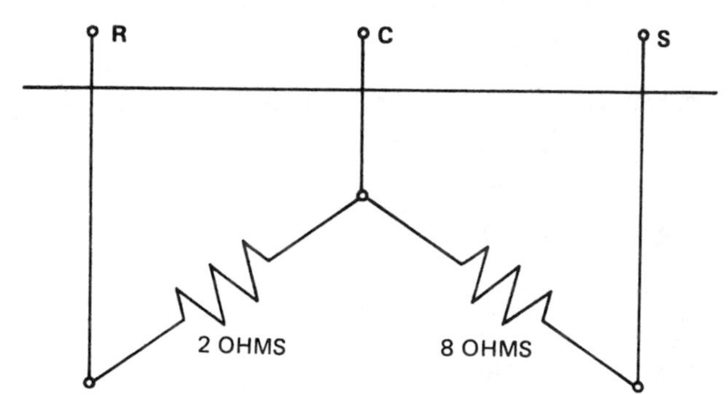

16. The left side, right side, and back panel heaters of a refrigerator are wired in parallel. Find the equivalent resistance of the circuit. __________

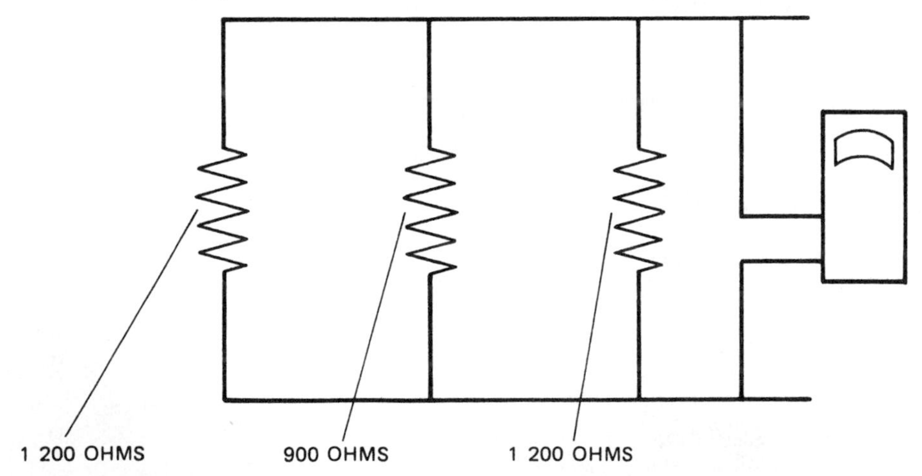

17. Two resistors are in a parallel circuit. The resistors have resistances of 18 ohms and 24 ohms. What is the equivalent resistance of the circuit? __________

18. Two electric baseboard heater units are placed in a room in series. The resistances of the units are 97 ohms and 291 ohms. What is the total resistance for the room? __________

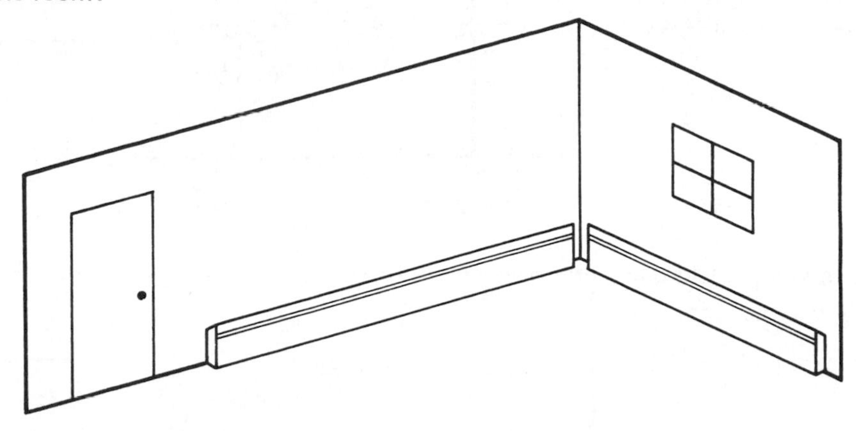

Note: Use this information for problems 19 and 20.
Some electrical circuits contain capacitors. A *capacitor* consists of two plates of electrical conducting material separated by an insulating material. *Capacitance* is the amount of electric charge a capacitor receives for each volt of applied potential. The unit of measure for capacitance is *microfarads*. For capacitors wired in parallel, the total capacitance is the sum of each capacitor's capacitance.

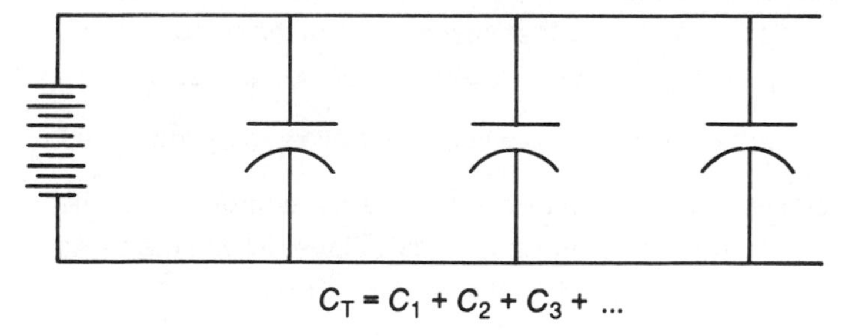

$$C_T = C_1 + C_2 + C_3 + \ldots$$

For capacitors wired in series, the equivalent capacitance is:

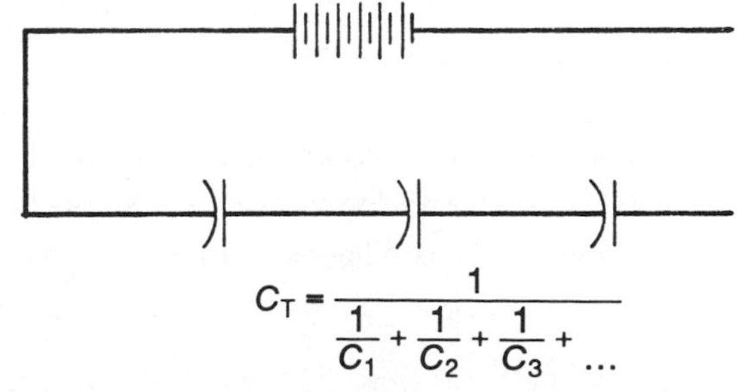

$$C_T = \frac{1}{\frac{1}{C_1} + \frac{1}{C_2} + \frac{1}{C_3} + \ldots}$$

Note: This formula is solved in the same way that $R_T = \frac{1}{\frac{1}{R_1} + \frac{1}{R_2} + \frac{1}{R_3} + \ldots}$ was.

19. A repairer must replace a faulty capacitor. The capacitor is replaced with two capacitors wired in parallel. What is the total capacitance? ____________

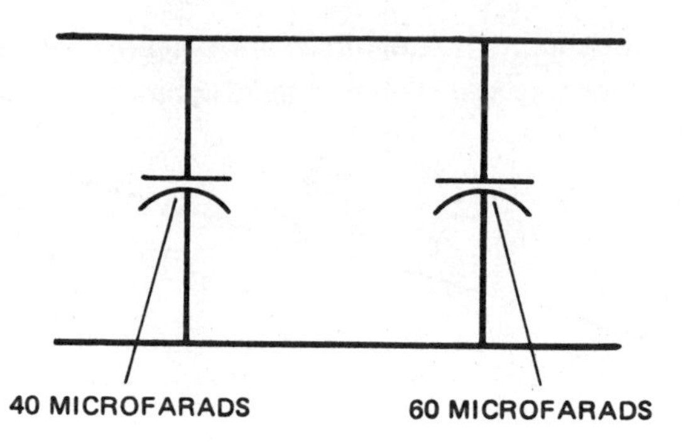

20. These relay capacitors are wired in series. Find, in microfarads, the total capacitance. ____________

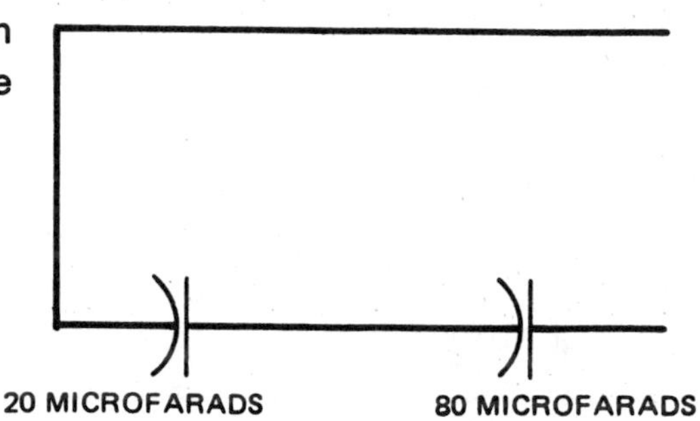

21. Wires act as resistance in circuits. This causes the voltage to decrease at the end of the wire. A building code states that a device should not be wired into a circuit if the voltage drop in the wires to that device is 5% of the rated voltage. A central air conditioning system is to be wired into a 230 volt system.

    a. What is the largest voltage drop in the wires that is allowed by the code? a. ____________

    b. If the starting current of the compressor is 0.089 ampere, find the resistance of the wire that would produce a 5% voltage drop in the wires to the compressor motor. Round off to the nearer tenth ohm. b. ____________

    c. If *#10* electrical cable has a resistance of $0.448 \frac{\text{ohm}}{\text{foot}}$, what is the maximum length, in feet, the cable can be without producing the 5% voltage drop at the compressor? Round off to the nearer whole foot. c. ____________

22. Can two 25 000-Btu room air conditioners be wired into the same electrical circuit protected by a 25-amp fuse? Each air conditioner is rated at 3 050 watts for a 230 volt circuit. (This rating is while the air conditioner is running steadily.) ____________

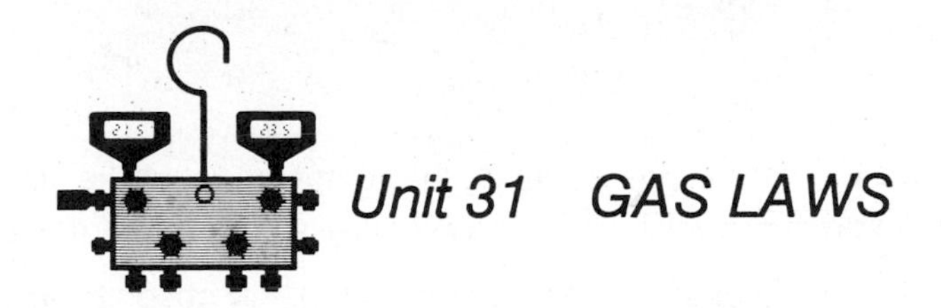

Unit 31 GAS LAWS

BASIC PRINCIPLES

- Study and apply these principles of gas laws to the problems in this unit.

This unit deals with formulas relating to the properties of gases. They are handled as the formulas were handled in the last unit.

Fluid pressure is the force that a gas or a liquid exerts per unit area. It is expressed in pounds per square inch (psi).

The pressure of a gas is usually measured by using a pressure gauge. A pressure gauge usually reads zero under normal atmospheric conditions. A gauge pressure reading is given as pounds per square inch gauge reading (psig).

The air also exerts a pressure. This pressure is called *atmospheric pressure*. Atmospheric pressure at sea level is 14.7 psi.

Absolute pressure uses both the gauge pressure reading and the atmospheric pressure. It is measured in pounds per square inch absolute (psia).

$$\text{absolute pressure} = \text{gauge pressure} + 14.7 \text{ psi}$$

The pressure of a gas is dependent upon temperature and the volume of the gas. A change in one of these three will result in a change in one or both of the others.

If the temperature of a gas remains constant, a change in the pressure of the gas is indirectly, or inversely, proportional to the volume of the gas. This relationship is known as *Boyle's Law*.

$$P_1 \times V_1 = P_2 \times V_2$$

where P_1 is the original absolute pressure
V_1 is the original volume
P_2 is the new absolute pressure
V_2 is the new volume

This means that if the pressure of the gas doubles, the new volume is one-half the original volume. If the new volume is four times the original volume, the new pressure is one-fourth the original pressure.

If the volume of the gas remains constant, a change in the pressure of the gas is directly proportional to the temperature of the gas. This relationship is known as *Charles' Law*.

$$\frac{P_1}{T_1} = \frac{P_2}{T_2}$$

or

$$P_1 \times T_2 = P_2 \times T_1$$

where P_1 is the original absolute pressure
T_1 is the original absolute scale temperature
P_2 is the new absolute pressure
T_2 is the new absolute scale temperature

This means that if the pressure of the gas doubles, the temperature of the gas doubles. If the new temperature is four times the original temperature, the new pressure is four times the original pressure.

Temperatures are measured on the absolute temperature scale. An absolute scale temperature is the Celsius scale temperature plus 273°. Absolute temperature is measured in kelvins (K).

$$K = °C + 273 \quad \text{or} \quad K = \left(\frac{5}{9} \times °F\right) + 255\frac{2}{9}$$

When neither the pressure, nor the temperature, nor the volume remains constant, the relationship is known as the *general law of perfect gas*.

$$\frac{P_1 \times V_1}{T_1} = \frac{P_2 \times V_2}{T_2}$$

or

$$P_1 \times V_1 \times T_2 = P_2 \times V_2 \times T_1$$

- Study this table of formulas for finding original and new pressures, temperatures, and volumes.

GAS LAW	ORIGINAL VALUE	NEW VALUE
Boyle's Law $P_1 \times V_1 = P_2 \times V_2$	$P_1 = \frac{P_2 \times V_2}{V_1}$	$P_2 = \frac{P_1 \times V_1}{V_2}$
	$V_1 = \frac{P_2 \times V_2}{P_1}$	$V_2 = \frac{P_1 \times V_1}{P_2}$
Charles' Law $\frac{P_1}{T_1} = \frac{P_2}{T_2}$ *or* $P_1 \times T_2 = P_2 \times T_1$	$P_1 = \frac{P_2 \times T_1}{T_2}$	$P_2 = \frac{P_1 \times T_2}{T_1}$
	$T_1 = \frac{P_1 \times T_2}{P_2}$	$T_2 = \frac{P_2 \times T_1}{P_1}$
General Law of Perfect Gas $\frac{P_1 \times V_1}{T_1} = \frac{P_2 \times V_2}{T_2}$ *or* $P_1 \times V_1 \times T_2 = P_2 \times V_2 \times T_1$	$P_1 = \frac{P_2 \times V_2 \times T_1}{V_1 \times T_2}$	$P_2 = \frac{P_1 \times V_1 \times T_2}{V_2 \times T_1}$
	$V_1 = \frac{P_2 \times V_2 \times T_1}{P_1 \times T_2}$	$V_2 = \frac{P_1 \times V_1 \times T_2}{P_2 \times T_1}$
	$T_1 = \frac{P_1 \times V_1 \times T_2}{P_2 \times V_2}$	$T_2 = \frac{P_2 \times V_2 \times T_1}{P_1 \times V_1}$

Be sure that *P* and *V* both are in absolute units. In other words use psia and K, not psig or °F or °C.

Note: Always use the formula that has what you are looking for by itself on the left side of the equal sign.

PRACTICAL PROBLEMS

Round the answer to the nearer hundredth when necessary.

Complete this chart.

	P_1	V_1	T_1	P_2	V_2	T_2
1.	30 psia	20 cu in	constant	50 psia	___ cu in	constant
2.	25 psia	14 cu in	constant	___ psia	21 cu in	constant
3.	20 psia	constant	27° C	___ psia	constant	77° C
4.	10 psia	constant	17° C	15 psia	constant	___ °C
5.	12 psia	constant	77° C	___ psia	constant	59° C
6.	24 psia	2 cu in	27° C	___ psia	1.3 cu in	52° C
7.	18 psia	5 cu in	22° C	12 psia	4.5 cu in	___ °C

8. An air compressor begins its cycle with 0.8 cubic inch of air at atmospheric pressure (14.7 psi or 0 psig) in its cylinder. The air leaving the cylinder has an absolute pressure of 42 psia. The temperature remains the same. What is the new volume of the air leaving the compressor? ______

9. A compressor takes 1.2 cubic inches of gas at a pressure of 14.7 psia and compresses it into a volume of 0.36 cubic inch. If the temperature of the gas remained the same, find the pressure of the gas after it is compressed. ______

10. A tank of gas has a temperature of 77°C. A gauge on the tank reads 105 psig. The tank cools to 27°C. What is the new gauge reading? ______

11. The original pressure of a container of gas is 80 psia. The original temperature is 207°C. The pressure is decreased to 50 psia. Find, in degrees Celsius, the new temperature of the gas. ______

12. In the morning, a tank containing refrigerant R-12 is at a pressure of 122.4 psia. The temperature is 50°F. In the afternoon, the temperature is 95°F. What is the new pressure inside the tank? ______

13. Find the exhaust pressure for the compressor. ____________

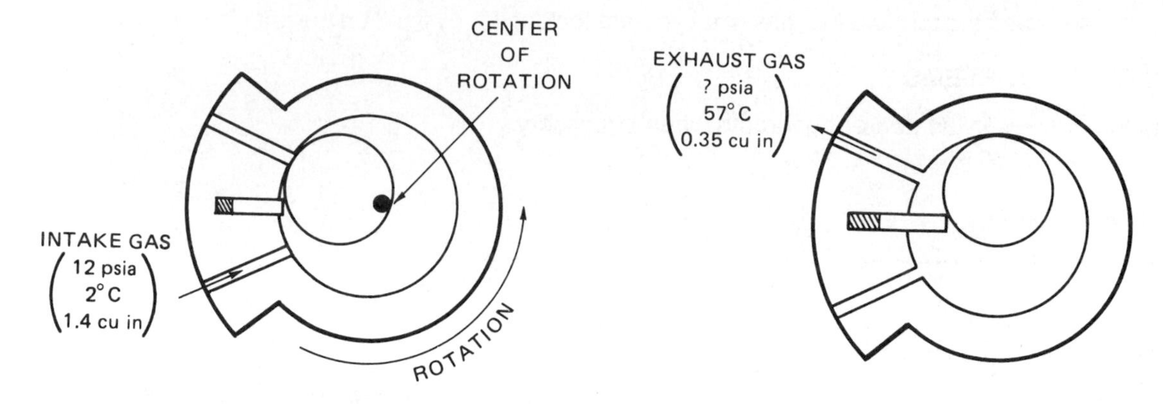

Note: Use this diagram for problems 14 and 15.

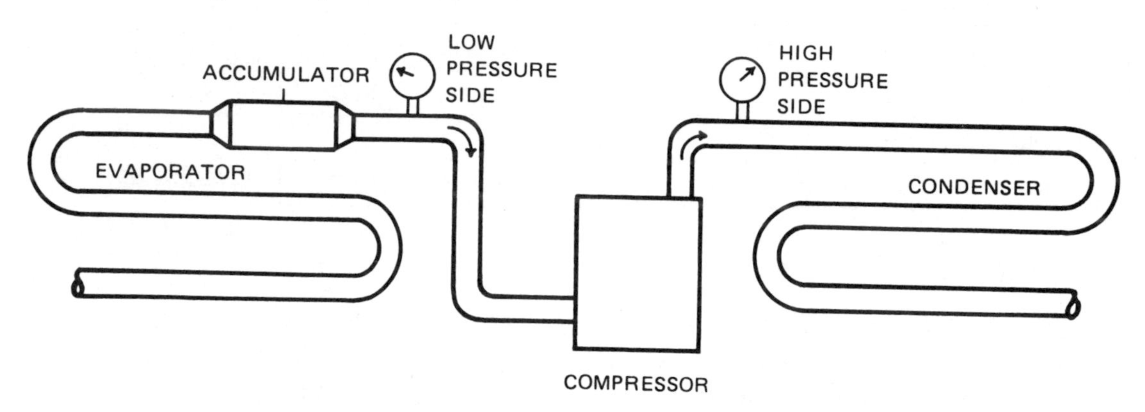

14. The values for the vaporator and condenser side of the air conditioner compressor system are given. Find, in kelvins, the temperature of the condenser side. ____________

EVAPORATOR SIDE	CONDENSER SIDE
P = 12.5 psia	P = 174 psia
V = 1.2 cu in	V = 0.11 cu in
T = 245 K	T = ?

15. The values for the low pressure and high pressure side of this air conditioner compressor system are given. Find the missing value. Round the answer to the nearer hundredth. ____________

LOW PRESSURE SIDE	HIGH PRESSURE SIDE
P = 0.3 psig	P = 137.3 psig
V = 0.95 cu in	V = ?
T = 3° C	T = 47° C

Unit 32 HEAT LOAD CALCULATIONS

BASIC PRINCIPLES

- Review and apply the principles of formula manipulation to these problems.

One activity that a heating and cooling technician may have to perform is determining the size of a heating or cooling system. Too small a system will make the customer unhappy because the house is too hot or too cool, and too large a system will cost more than a proper system would cost. It is important to size the unit correctly. To get the correct size unit, the proper heat load must be determined.

The heat load is the amount of heat which would be lost or gained each hour with the extreme temperature difference for which the system is designed. Heat loads are not calculated for interior structures since no heat is lost through these structures. Heat loads are determined by multiplying the heat transfer multiplier (the amount of heat transferred through one square foot of the structure) for a type of structure by the area of that structure. The formula is written as

$$\text{heat load} = \text{heat transfer multiplier} \times \text{area}$$

See the appendix or page 158 for the table of Heat Transfer Multipliers.

Note: Always use the formula that has what you are looking for by itself on the left side of the equals sign.

PRACTICAL PROBLEMS

Express the answer to the nearer British thermal unit per hour.

1. A house is built in a location with a design temperature difference of 75°F. The room shown is a corner room on the second story, so it loses heat only through the two outside walls and the ceiling. The walls are wood frame with sheathing and siding and 3 1/2 inches of insulation (R-11). The windows are single pane. The ceiling has 6 inches of insulation (R-19) under a vented attic. Find the heat load for this room. __________

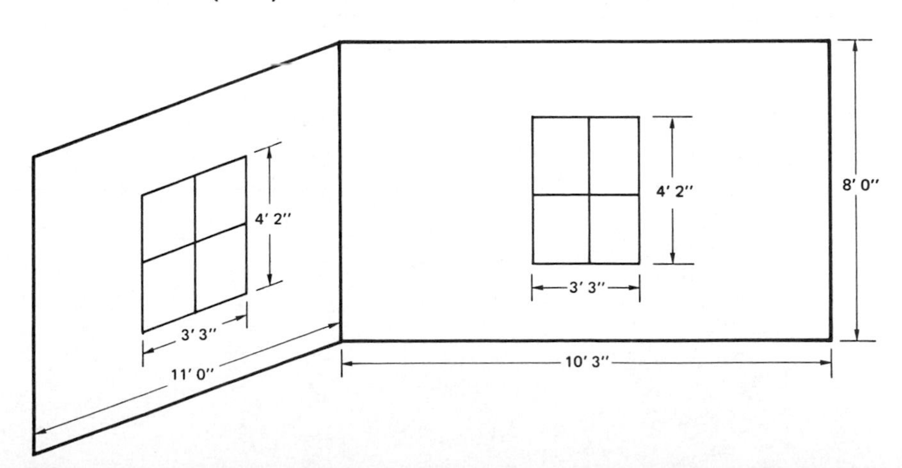

HEAT TRANSFER MULTIPLIERS

Note: The Heat Transfer Multiplier is found by multiplying the U factor (the amount of heat transferred through 1 square foot of structure for each degree temperature difference between the inside and outside surfaces) by the design temperature difference. The units for the Heat Transfer Multiplier are British thermal units per hour per square foot.

TYPE OF STRUCTURE	DESIGN TEMPERATURE DIFFERENCE		
	25°F	70°F	75°F
Walls — wood frame with sheathing and siding or other veneer			
3 1/2 inches insulation (R-11)	3.5	5	5
3 1/2 inches insulation + 1 inch polystyrene sheathing	3.1	3.5	3.8
Ceiling — under vented roof			
3 1/2 inches insulation (R-11)	2.5	6	6
6 inches insulation (R-19)	1.5	4	4
9 1/2 inches insulation (R-30)	1.0	2.2	2.4
Floor			
No insulation	5	16	17
6 inches insulation	1	3.2	3.4
Windows			
Single pane	35	105	110
Double pane	25	70	75
Single pane + storm window	25	60	65
Double pane (fixed)	25	60	65
Doors			
Insulated core, weather-stripped	5.3	81	86
Sliding glass door, double glass	25	90	95

2. A one-story house is built over a vented crawl space. The attic is also vented. The heating system is designed for a temperature difference of 75°F. The exterior wall of a middle bedroom is shown. The room is 10 feet 3 inches wide. The window has double pane glass. The wall is wood frame with sheathing and siding. It has 1 inch of polystyrene insulation over 3 1/2 inches of insulation (total R-16). The ceiling has 9 1/2 inches of insulation (R-30), and the floor has 6 inches of insulation (R-19). Find the heat load for this room. __________

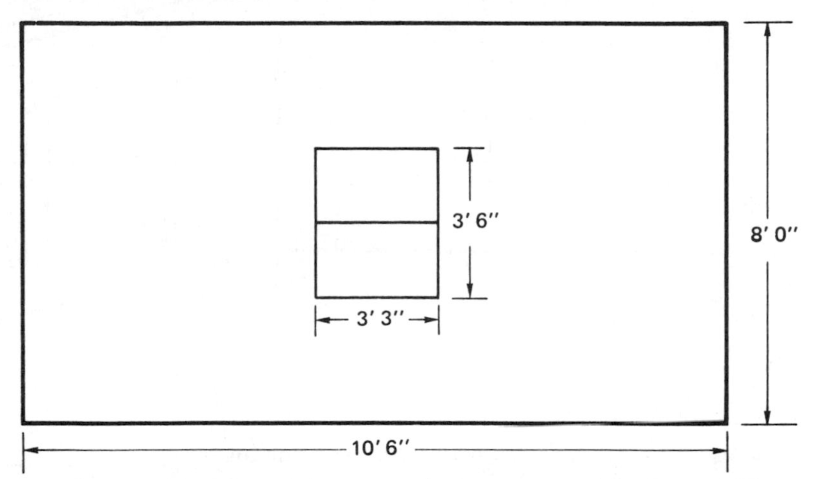

3. A dining room of a two-story house is located in the corner of the house. The house is designed for a temperature difference of 70°F. The dining room has a sliding glass door with double glass. The walls are frame with 3 1/2 inches of insulation (R-11) and sheathing and siding. The floor has 6 inches of insulation (R-19), and is over an unheated basement. This basement gives the floor a design temperature difference of 25°F.

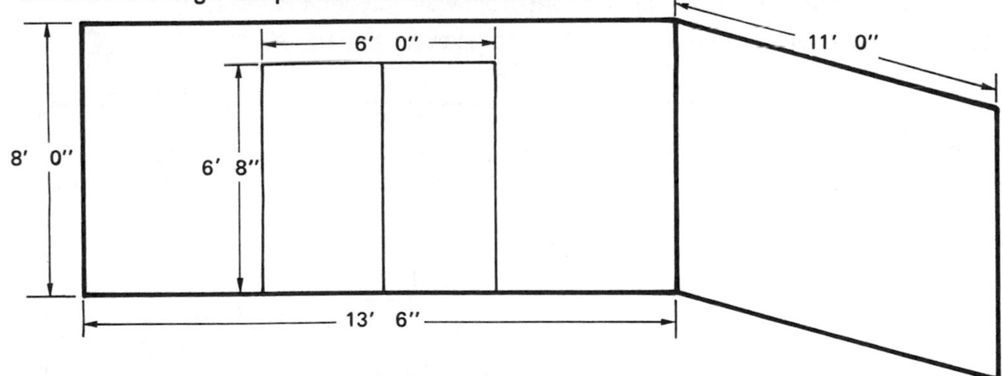

   a. What is the heat load for this room? a. __________
   b. What percent of the heat is lost through the sliding glass door? Express the answer to the nearer tenth percent. b. __________

4. A corner bedroom is situated on the second story of a two-story house. The house is designed for a temperature difference of 70°F. Above the bedroom is a vented attic. There are 9 1/2 inches of insulation (R-30) on the ceiling. The walls are frame with sheathing and siding and have 1 inch of polystyrene insulation over 3 1/2 inches of insulation (total R-16) in them. The windows are single pane.

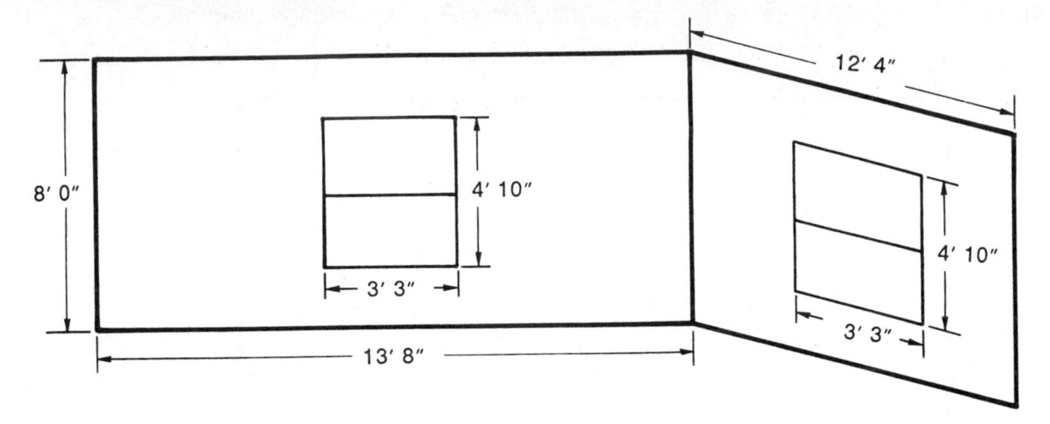

a. Find the heat load for this room. a. ____________

b. Find the heat load for the room if storm windows were installed. b. ____________

5. A house is designed for a temperature difference of 75°F. The living room has an insulated core door that is weather-stripped. The windows are double pane with the center section fixed. This living room is on the first floor of a two-story house and is over a vented crawl space. The floor has 6 inches of insulation (R-19) in it. The side wall is frame and has 3 1/2 inches of insulation plus 1 inch of polystyrene insulation (total R-16) on it. The front wall has brick veneer over the same type of insulation (R-16). Find the heat load for this room. ____________

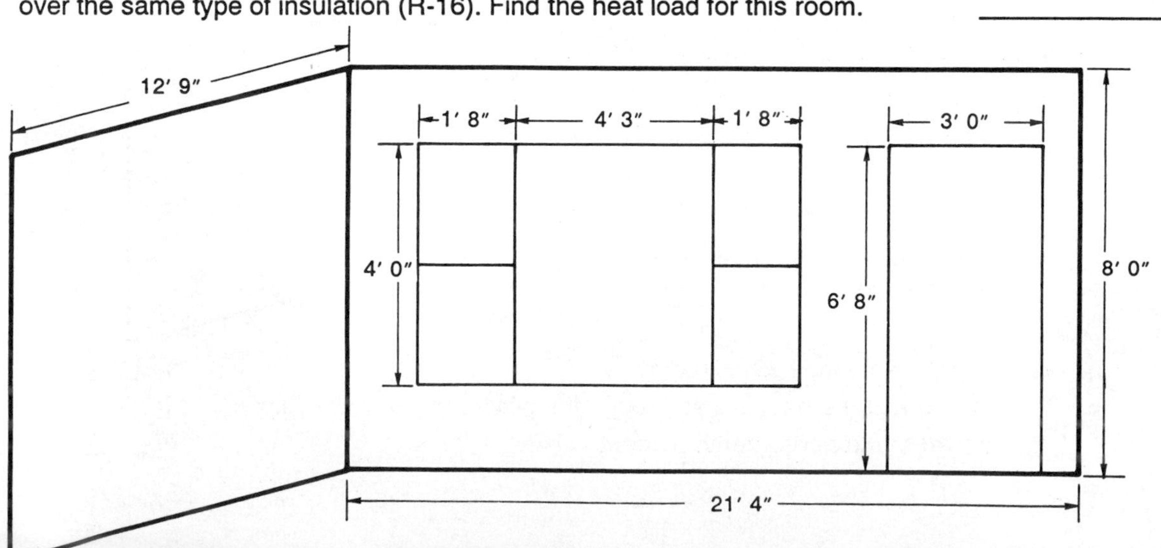

6. A one-story house built over a vented crawl space is designed for a 70°F temperature difference. The attic area is vented. The walls all have 3 1/2 inches of insulation (R-11) with sheathing and siding. The ceilings have 9 1/2 inches of insulation (R-30), and the floor has 6 inches (R-19). All windows are double pane and each one is 4 feet 2 inches high and 3 feet 3 inches wide. The sliding glass door is 6 feet 8 inches high and 6 feet wide. Both doors are 6 feet 8 inches high and 3 feet wide, have insulated cores, and are weather-stripped. The ceilings are 8 feet high. Find the heat load for the house. ______________

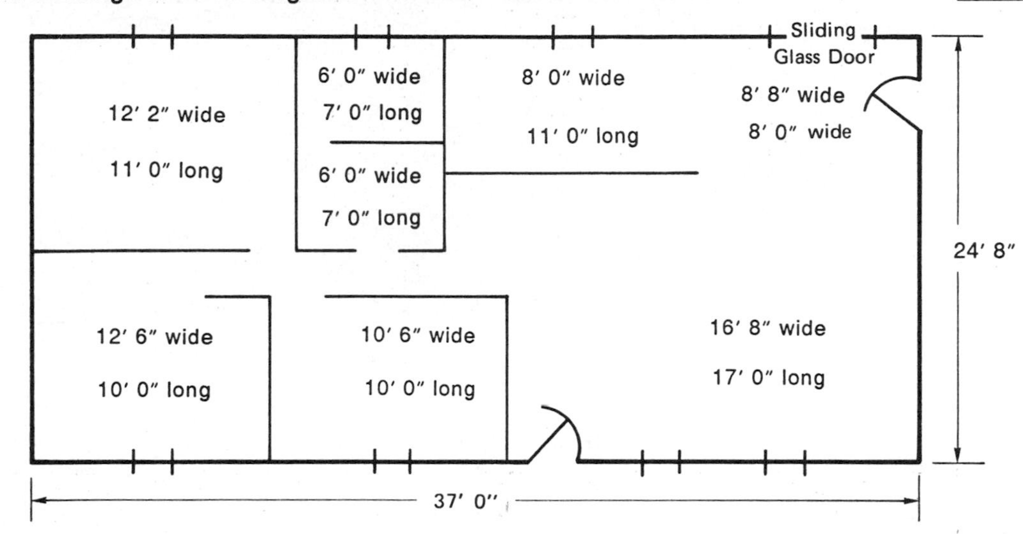

7.

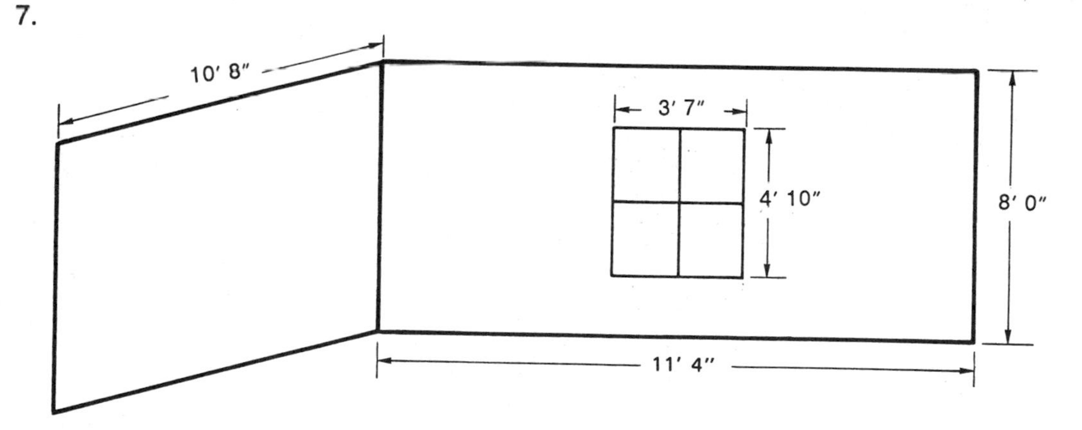

a. The cooling load for a house is determined in part by using the same formula used to find the heating load. The temperature difference used in designing a cooling system is 25°F. The room shown does not face the sun so no additional cooling load is required. The windows are single pane and the walls are brick veneer with 3 1/2 inches of insulation (R-11).

The ceiling has 3 1/2 inches of insulation (R-11), and the floor has no insulation. The room is built over a vented crawl space with a vented attic above it. Find the cooling load for this room. a. ____________

b. If the room faced the sun, an additional cooling load would be created. This extra cooling load would be due to the sun shining through the window and heating the room. A window facing south would allow a maximum of 75 Btu per hour to enter the room for each square foot of window area. If the room has its window face south, what is the maximum cooling load for the room? b. ____________

Stretchouts and Lengths of Arcs

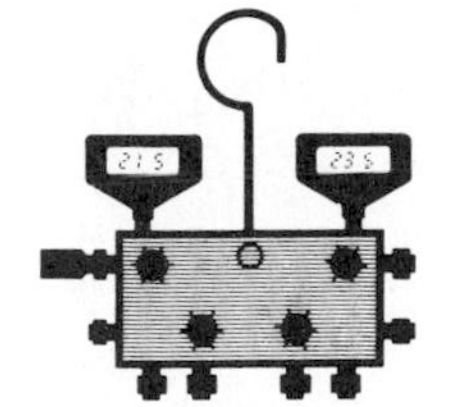

Unit 33 STRETCHOUTS OF SQUARE AND RECTANGULAR DUCTS

BASIC PRINCIPLES

- Study and apply these principles of stretchouts of square and rectangular ducts to the problems in this unit.

There are times when a heating and cooling technician must handcraft a duct. It can be done by trial and error or by guesswork, but that can be a waste of material that could be costly. Cutting out the duct correctly the first time is efficient and cost effective.

In finding the amount of material needed for a square or rectangular duct, the size of the duct pattern, or the *stretchout*, is needed.

The length of the stretchout (*L.S.*) is the perimeter of the end of the duct plus the allowance for the seam (*W*).

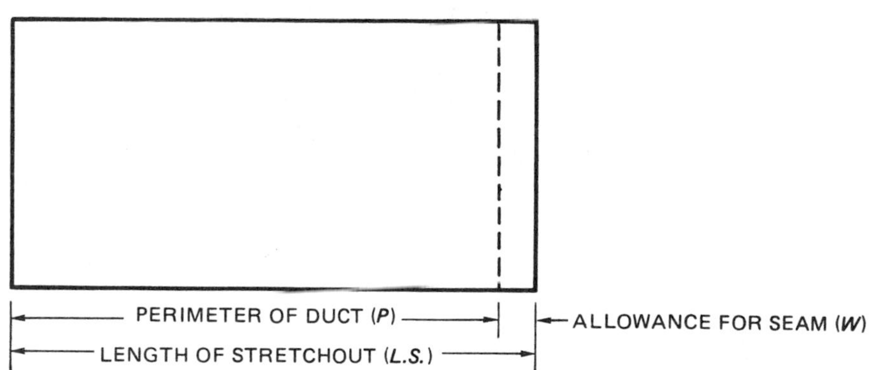

The dimensions of the duct, the width of the seam, and the type of seam will determine the actual layout for the stretchout.

The width of the stretchout (*W.S.*) is the length of the duct (*l*) plus the allowance for the edge or seam (*L*).

LENGTH OF DUCT (l)
WIDTH OF STRETCHOUT (*W.S.*)
ALLOWANCE FOR EDGE (*L*)

Some ducts will have end seams on both ends. Some ducts will not have end seams. The actual layout for the stretchout will depend upon the edges and end seams.

Note: In finding the stretchout for square and rectangular ducts, the thickness of the metal is not considered.
The key to choosing the correct formula is determining the type of seam on the duct.
The length of the seam will always be dimension I.

PRACTICAL PROBLEMS

Note: Use this information for problems 1–3.

BUTT OR WELDED SEAM DUCTS

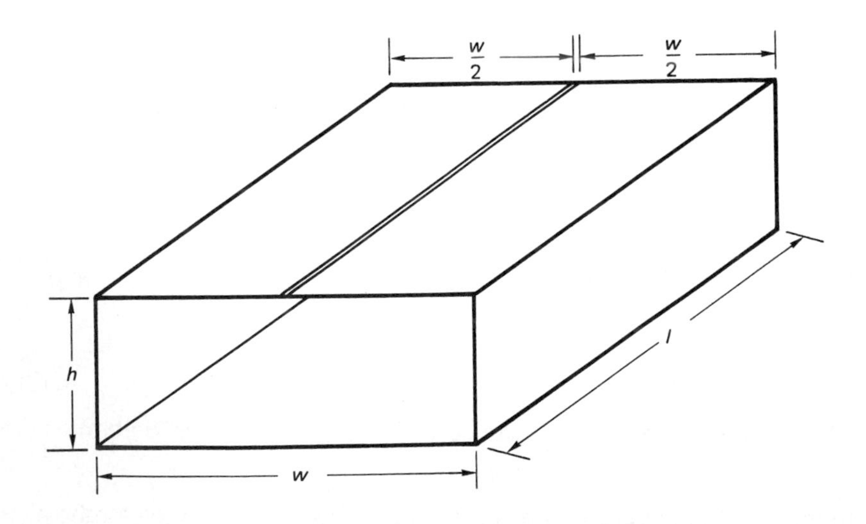

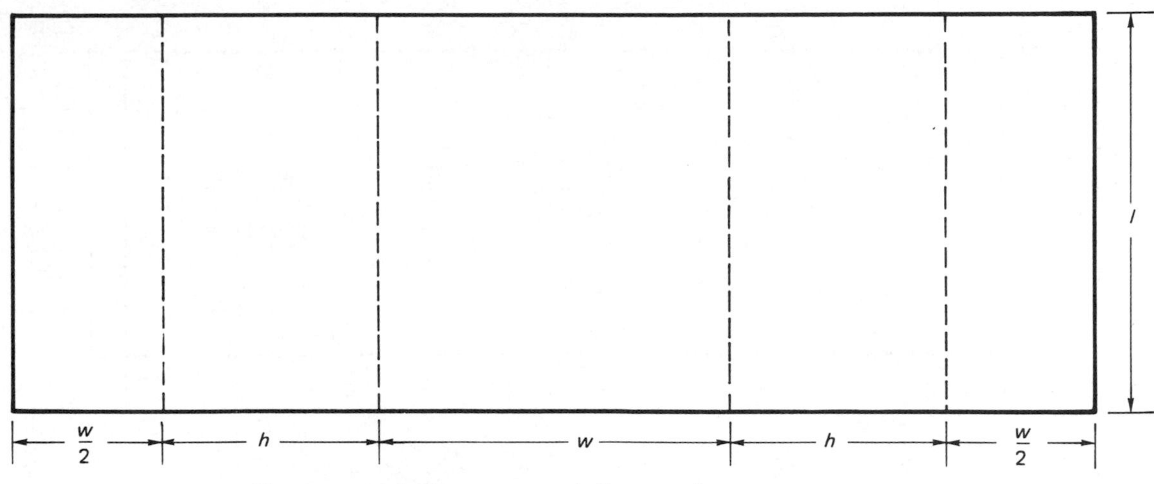

For square ducts: $L.S. = 4s$

For rectangular ducts: $L.S. = 2h + 2w$

1. A 12-inch square duct is 18 inches long.
   a. Find the length of the stretchout. a. ______
   b. Find the width of the stretchout. b. ______
2. A 3-foot long rectangular duct is 8 inches wide and 10 inches high.
   a. What is the stretchout length in inches? a. ______
   b. What is the stretchout width in inches? b. ______
3. The dimensions of a rectangular duct are 6 inches wide by 10 inches high by 26 1/4 inches long. It has one end seam of 1/2 inch.
   a. Find the length of the stretchout. a. ______
   b. Find the width of the stretchout. b. ______

Note: Use this information for problems 4–7.

LAP SEAM DUCTS

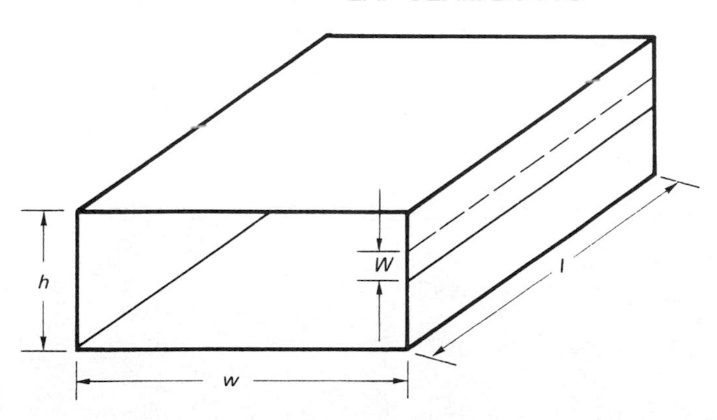

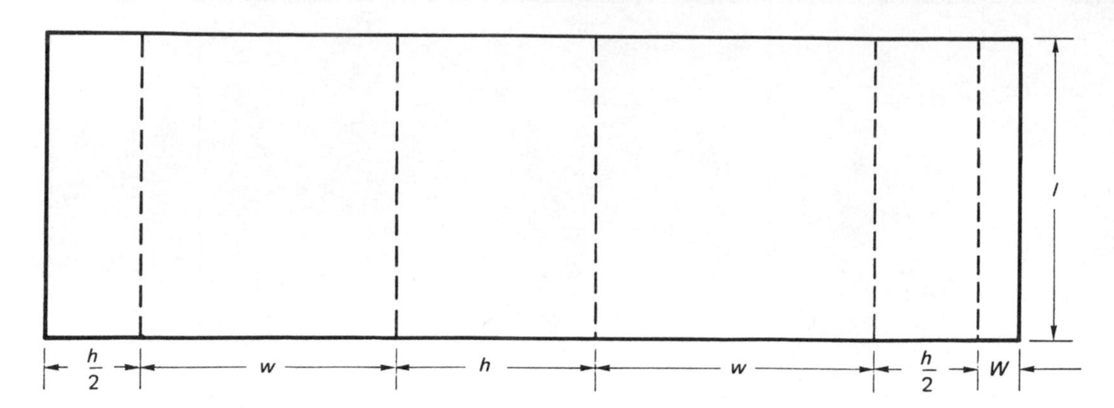

For square ducts: $L.S. = 4s + W$

For rectangular ducts: $L.S. = 2h + 2w + W$

4. A 2-foot long rectangular duct is 8 inches wide by 12 inches high. The lap seam is 3/8 inch.
   a. Find the length of the stretchout. a. ______________
   b. Find the width of the stretchout. b. ______________

5. The dimensions of a rectangular duct are:

   $h = 25$ cm; $w = 20$ cm; $l = 75$ cm; $W = 0.8$ cm
   a. What is the length of the stretchout in centimeters? a. ______________
   b. What is the width of the stretchout in centimeters? b. ______________

6. An 8-inch square duct has a lap seam of 1/4 inch. The duct is 30 inches long.
   a. Find the length of the stretchout. a. ______________
   b. Find the width of the stretchout. b. ______________

7. A rectangular duct is 2 feet wide, 30 inches high, and 2 feet long. The lap seam is 1/4 inch. The end seam is 3/4 inch.
   a. What is the length of the stretchout in inches? a. ______________
   b. What is the width of the stretchout in inches? b. ______________

Note: Use this information for problems 8–12.

GROOVED SEAM DUCTS

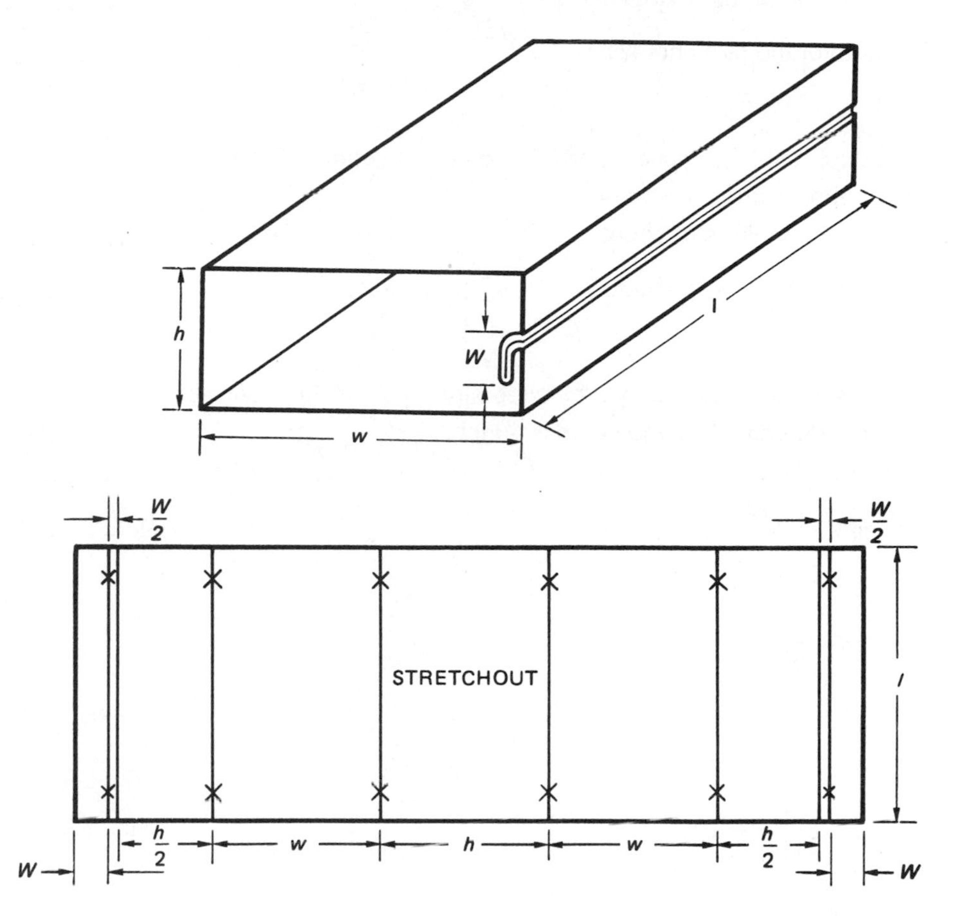

For square ducts: $L.S. = 4s + 3W$

For rectangular ducts: $L.S. = 2h + 2w + 3W$

8. A square duct measures 22 centimeters on each side. The duct is 1 meter long. It has a 0.75-centimeter grooved seam.
   a. Find, in centimeters, the length of the stretchout. a. ______
   b. Find, in centimeters, the width of the stretchout. b. ______

9. The dimensions of a square duct are:

   s = 9 inches; l = 3 feet; W = 3/8 inch

   The seam on the duct is grooved.
   a. Find, in inches, the value of *L.S.* a. ______
   b. Find, in inches, the value of *W.S.* b. ______

10. The width of the grooved seam on a rectangular duct is 1/4 inch. The duct is 6 inches wide by 12 inches high by 32 1/2 inches long.
    a. What is the length of the stretchout? a. ____________
    b. What is the width of the stretchout? b. ____________

11. A 30-inch square duct is 27 1/2 inches long. The grooved seam is 3/8 inch. There is a 1/4-inch end seam on both ends.
    a. Find the total length of the stretchout. a. ____________
    b. Find the total width of the stretchout. b. ____________

12. A rectangular duct is 1 1/2 feet wide by 30 inches high by 22 inches long. It has a 1/4-inch grooved seam. The end seam is 1 inch.
    a. What is the value of *L.S.*? a. ____________
    b. What is the value of *W.S.*? b. ____________

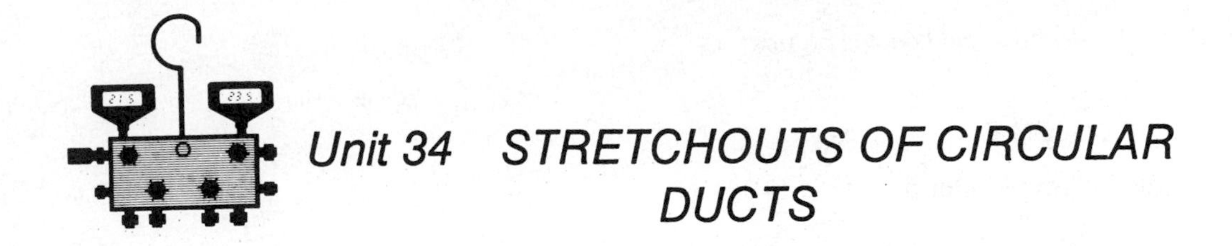

Unit 34 STRETCHOUTS OF CIRCULAR DUCTS

BASIC PRINCIPLES

- Study and apply these principles of stretchouts of circular ducts to the problems in this unit.

In finding the amount of material needed for a circular duct, the size of the duct pattern, or the *stretchout*, is needed.

Determining the stretchout for a circular duct is similar to finding the stretchout for a rectangular duct, except that a different formula is used.

The length of the stretchout (*L.S.*) is the circumference at the end of the duct (*C*) plus the allowance for the seam (*W*). The width of the stretchout (*W.S.*) for circular ducts without end and edge seams is the length of the duct (*l*).

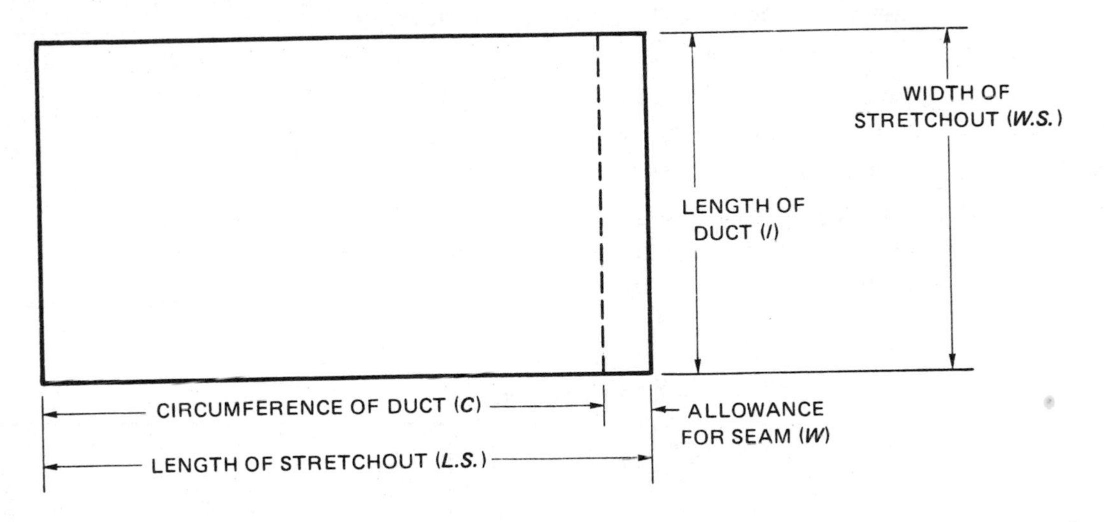

The dimensions of the duct, the width of the seam, and the type of seam will determine the actual layout for the stretchout.

Note: In finding the stretchout for circular ducts made with gauge metal, the thickness of the metal is not considered.

Review the tables of length equivalents in section II of the appendix.

Note: The key to choosing the correct formula is determining the type of seam on the duct. The length of the seam will always be dimension l.

PRACTICAL PROBLEMS

Note: Use this information for problems 1–3.

BUTT OF WELDED SEAM DUCTS

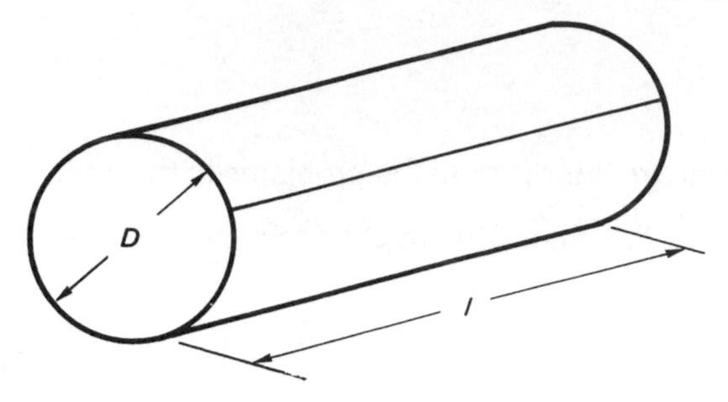

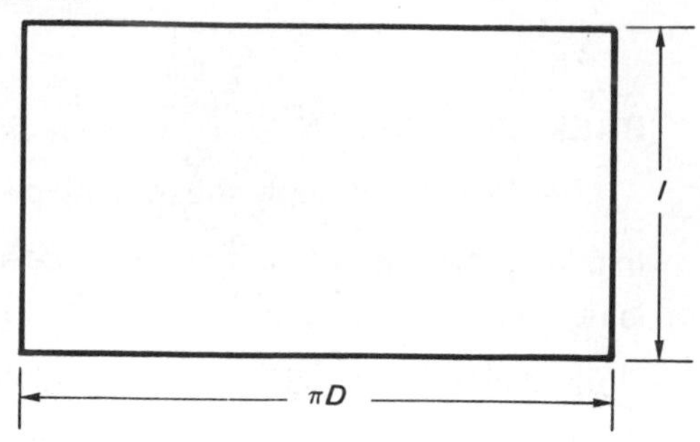

For circular ducts: $L.S. = 2\pi r$ or $L.S. = \pi D$

where $\pi = 3.141\,6$

1. The diameter of a circular duct is 6 inches. The duct is 24 inches long.
   a. Find, to the nearest 16th inch, the length of the stretchout. a. ____________
   b. Find the width of the stretchout. b. ____________
2. A circular duct is 20 inches long. The diameter of the duct is 8 1/2 inches.
   a. What is the length of the stretchout? Round the answer to the nearer 16th inch. a. ____________
   b. What is the width of the stretchout? b. ____________
3. A 24-centimeter diameter duct is 50 centimeters long.
   a. Find, to the nearer hundredth centimeter, the value of *L.S.* a. ____________
   b. Find, in centimeters, the value of *W.S.* b. ____________

Note: Use this information for problems 4–6.

LAP SEAM DUCTS

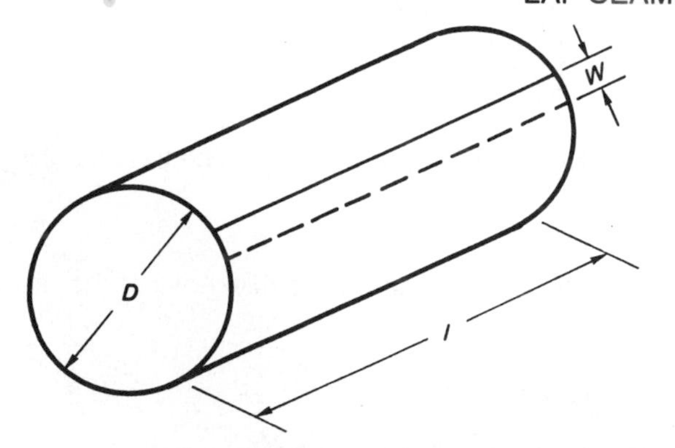

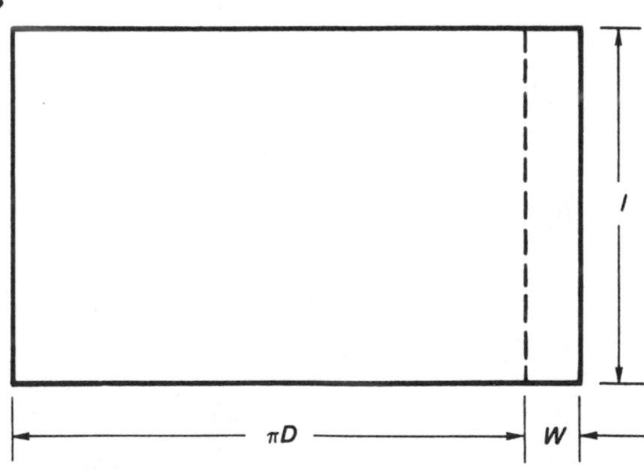

For circular ducts: $L.S. = 2\pi r + W$ or $L.S. = \pi D + W$

where $\pi = 3.141\,6$

4. The dimensions of a circular duct are:

$$D = 8 \text{ inches}; l = 3 \text{ feet}; W = 1/4 \text{ inch}$$

   a. What is the value of *L.S.* to the nearer 16th inch? a. ____________
   b. What is the value of *W.S.* in inches? b. ____________

5. The width of the lap seam on a circular duct is 0.7 centimeter. The duct is 1 meter long and has a diameter of 30 centimeters.
   a. Find, to the nearer hundredth centimeter, the length of the stretchout. a. ____________
   b. Find, in centimeters, the width of the stretchout. b. ____________

6. A 10-inch diameter duct has a length of 2 1/2 feet. The lap seam has a width of 3/8 inch.
   a. What is the length of the stretchout? Round the answer to the nearer 16th inch. a. ____________
   b. What is the width of the stretchout? b. ____________

Note: Use this information for problems 7–9.

GROOVED SEAM DUCTS

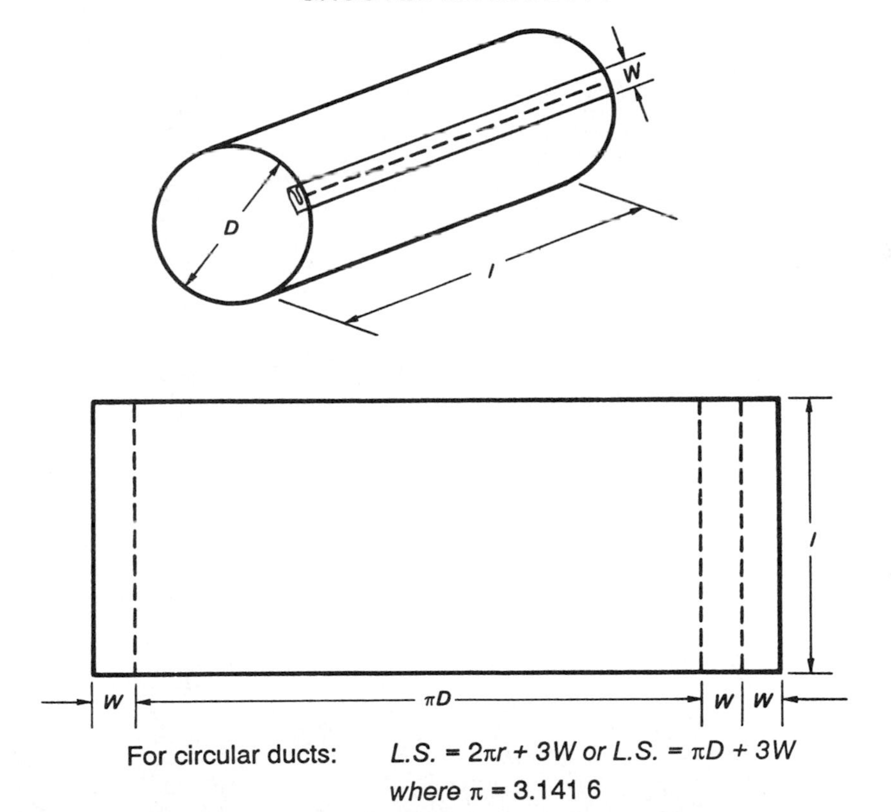

For circular ducts: $L.S. = 2\pi r + 3W$ or $L.S. = \pi D + 3W$ where $\pi = 3.141\,6$

7. The diameter of a circular duct is 7 1/2 inches. The length of the duct is 28 inches. The grooved seam on the duct has a width of 3/8 inch.
   a. How many inches long is the stretchout? Round the answer to the nearer 16th inch. a. ____________
   b. How many inches wide is the stretchout? b. ____________

8. A 9-inch diameter duct has a 1/4-inch wide grooved seam. The length of the duct is 2 feet.
   a. Find, to the nearer 16th inch, the length of the stretchout. a. ____________
   b. Find the width of the stretchout. b. ____________

9. The dimensions of a circular duct are:

$$D = 22 \text{ centimeters}; \; l = 70 \text{ centimeters}; \; W = 0.8 \text{ centimeter}$$

   a. What is the value of *L.S.* to the nearer hundredth centimeter? a. ____________
   b. What is the value of *W.S.* in centimeters? b. ____________

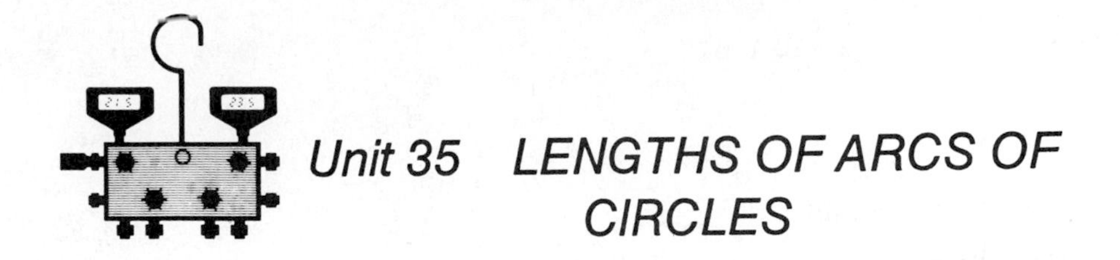

Unit 35 LENGTHS OF ARCS OF CIRCLES

BASIC PRINCIPLES

- Study and apply these principles of lengths of arcs of circles to the problems in this unit.

There are times when ducts must make turns. These ducts will have to be made from separate pieces. In order to properly cut out the pieces for the ducts, we need to study and become familiar with arcs of circles.

An arc is part of a circle. The number of degrees in an arc is measured by the central angle. The length of an arc (L) is a fraction times the circumference (C) of the circle.

The fraction is a ratio of the number of degrees in the arc ($n°$) to the number of degrees in the circle (360°).

$$L = \frac{n°}{360°} \times \pi D$$

$$L = \frac{n°}{360°} \times 2\pi r$$

where $\pi = 3.141\,6$

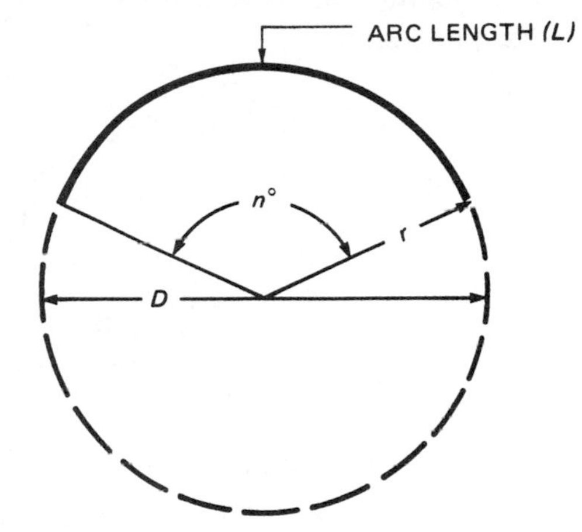

For a given arc length and circumference, the measure of the angle is:

$$n° = \frac{360° \times L}{\pi D}$$

or

$$n° = \frac{360° \times L}{2\pi r}$$

where $\pi = 3.141\,6$

Note: The radius of the heel of a rectangular elbow is not the width of the duct, but the distance from the center of the arc.

PRACTICAL PROBLEMS

1. Find, to the nearer hundredth inch, the length of this arc.

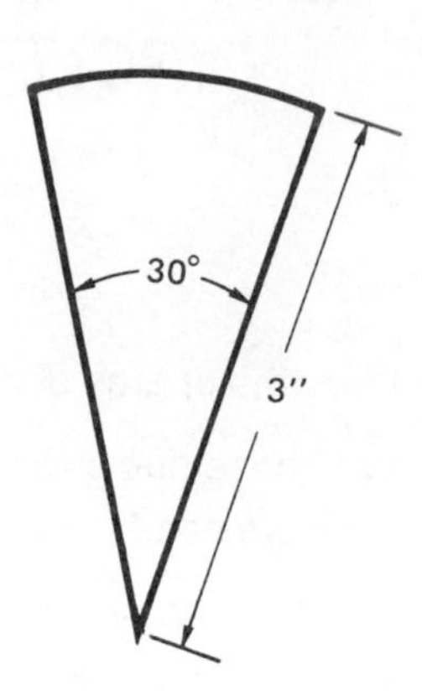

2. What is the length of this arc in centimeters? Round the answer to the nearer hundredth.

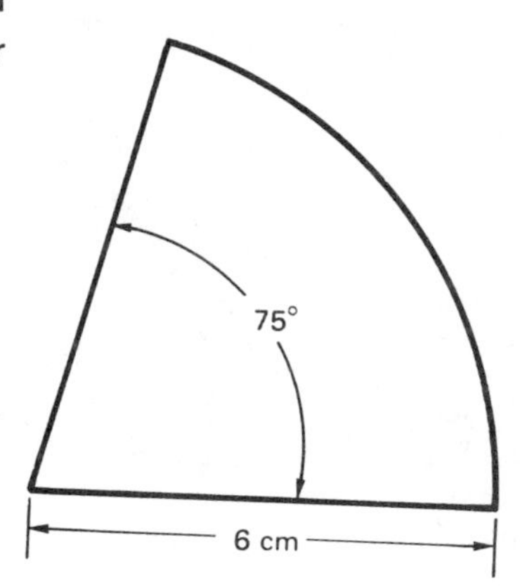

3. This arc has a radius of 4 feet. The central angle is 225°. How many feet are in the measure of the arc? Round the answer to the nearer hundredth.

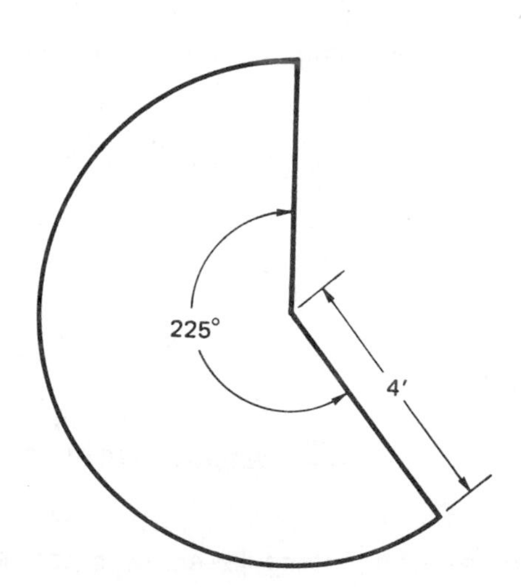

4. A 270° arc has a radius of 0.5 meter. Find, to the nearer hundredth meter, the length of the arc. __________

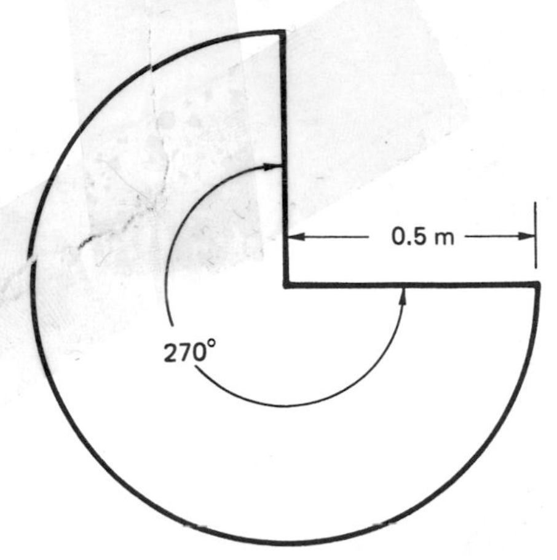

5. The length of an arc of a circle is 11.775 feet. The diameter of the circle is 9 feet. How many degrees are in the central angle of the arc? Round the answer to the nearer degree. __________

6. An arc has a length of 86.55 centimeters. The radius is 15 centimeters. Find the number of degrees in the central angle of the arc. Round the answer to the nearer degree. __________

7. The cylinder of a rotary compressor is 12 centimeters in diameter. The angle between the intake and exhaust ports of the compressor is 40°. What is the distance between the posts measured along the arc? Round the answer to the nearer hundredth centimeter. __________

8. What is the arc length of the center of this duct? Round the answer to the nearer hundredth inch. __________

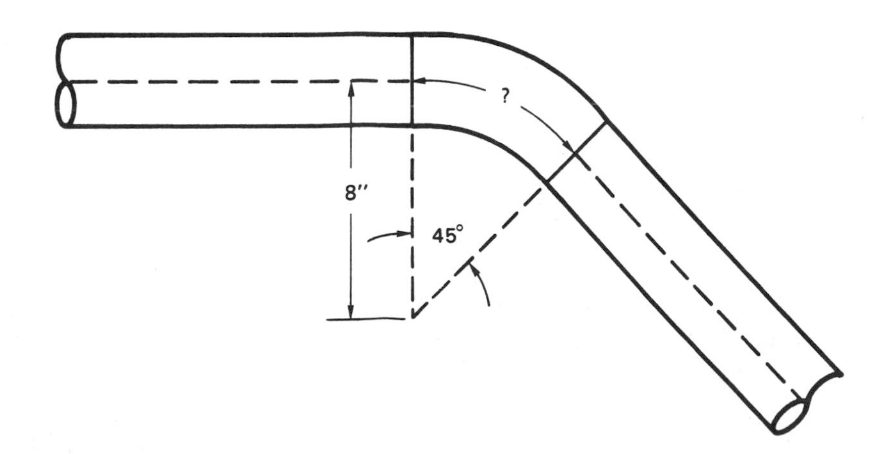

9. Openings in a gas furnace allow air to enter the burners. The openings of this furnace have a central arc length of 3/4 inch. Find, to the nearer degree, the angle of the opening. ______

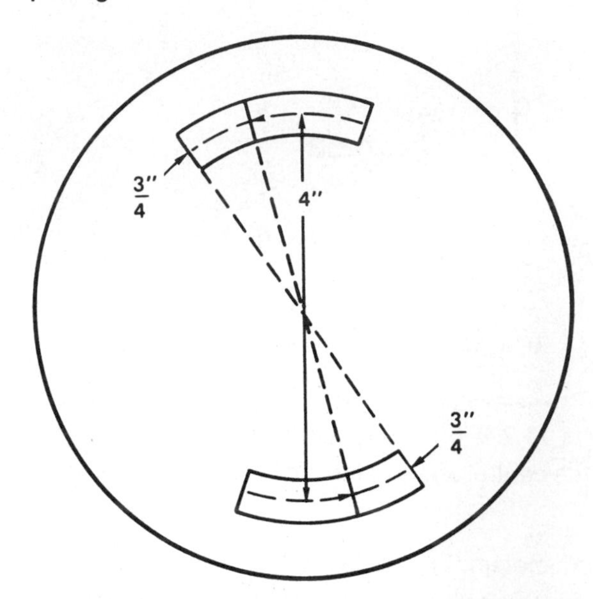

10. The control knob of a window air conditioning unit has four settings. The knob is turned from the OFF position to the HI COOL position. Find the number of degrees through which the knob is turned. Round the answer to the nearer whole degree. ______

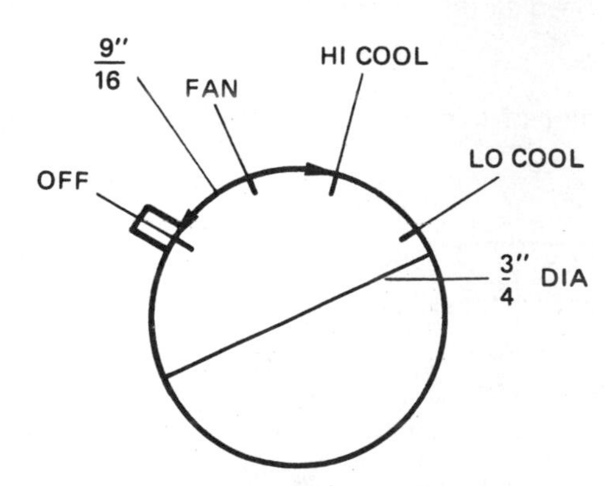

11. An oil gun is fastened to a furnace with 6 screws. The screws are equally spaced and form a 6-inch diameter circle. What is the arc length, to the nearer hundredth inch, between the centers of the screws? ______

12. There are 9 equally spaced markings along the edge of the circular temperature control dial of a refrigerator. The arc length between the markings is 2.7 centimeters.

    Using the formula $D = \dfrac{360° \times L}{n° \times 3.141\,6}$, find the diameter of the dial.

    Round the answer to the nearer hundredth centimeter. __________

Note: Use this information for problems 13–16.

A *rectangular 90° elbow* is an elbow with the heel and the throat at 90 degrees to the cheek, and the cheek is a 90° portion of a circle. A *rectangular 45° elbow* is an elbow with the heel and the throat at 90° to the cheek, and the cheek is a 45° portion of a circle.

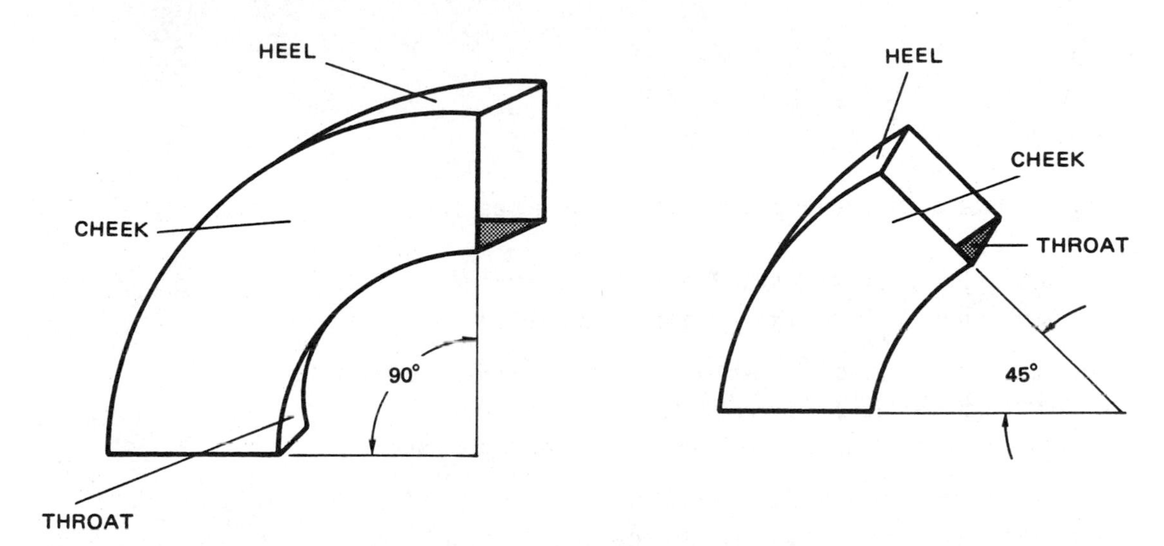

To find the length of the stretchout of the throat or the heel, the arc length is needed. Arc length is found by using the throat radius or the heel radius. The heel radius is the throat radius plus the width of the duct.

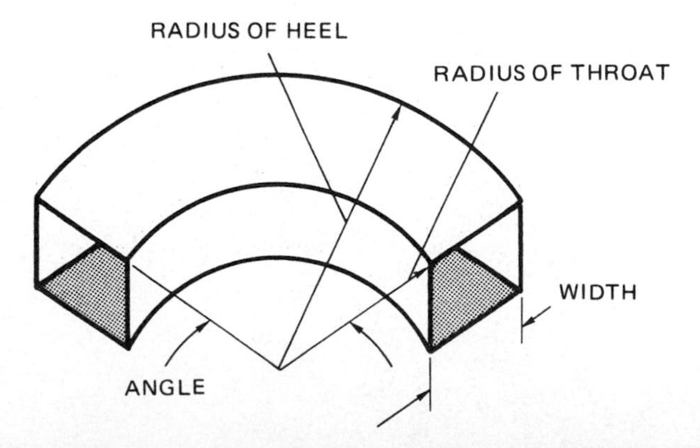

Note: The radius of the heel of a rectangular elbow is not the width of the duct, but the distance from the center of the arc.

13. The rectangular 45° elbow has a throat radius of 5 inches. The heel radius is 15 inches.
    a. What is the arc length of the throat to the nearer 16th inch?
    b. What is the arc length of the heel to the nearer 16th inch?

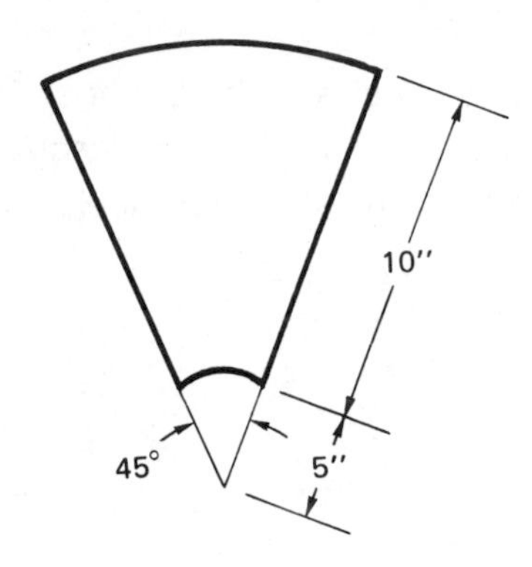

14. The throat radius of a rectangular 90° elbow is 5 inches. The duct is 10 inches wide.
    a. Find, to the nearer 16th inch, the length of the throat. a. ____________
    b. Find, to the nearer 16th inch, the length of the heel. b. ____________

15. A rectangular 45° duct is 9 inches wide. The radius of the throat is 4 1/2 inches.
    a. What is the arc length of the throat to the nearer 16th inch? a. ____________
    b. What is the arc length of the heel to the nearer 16th inch? b. ____________

16. A 30-centimeter wide duct has a throat radius of 15 centimeters. The angle between the sides of the heel and the cheek is 90°.
    a. Find, to the nearer hundredth centimeter, the arc length of the throat. a. ____________
    b. Find, to the nearer hundredth centimeter, the arc length of the heel. b. ____________

Trigonometry

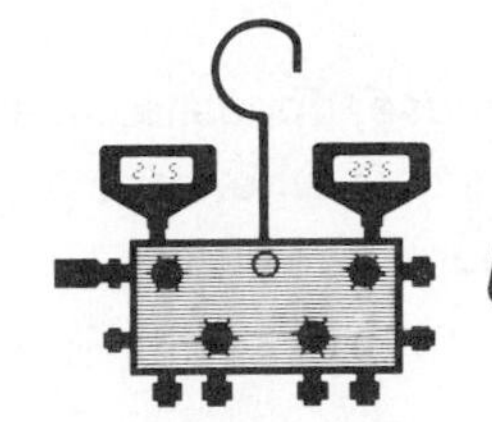

Unit 36 TRIGONOMETRIC FUNCTIONS

BASIC PRINCIPLES

- Study and apply these principles of trigonometric functions to the problems in this unit.

Triangles have the following property: when two triangles have the same angles, their sides will have the same ratios. This property can be used to determine angles or lengths of sides of triangles. The ratios of the sides of a right triangle are given special names and put in a table.

These six relationships are widely used in expressing the ratio of the sides of a right triangle.

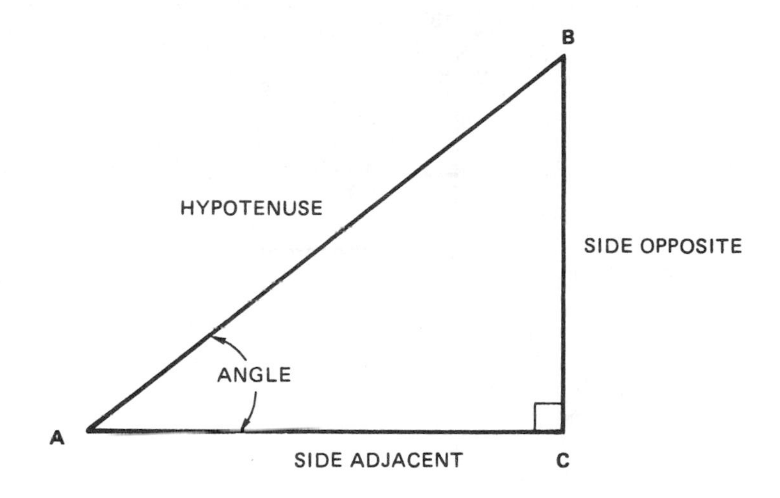

SINE

sin $= \dfrac{\text{side opposite}}{\text{hypotenuse}}$

COSINE

cos $= \dfrac{\text{side adjacent}}{\text{hypotenuse}}$

TANGENT

tan $= \dfrac{\text{side opposite}}{\text{side adjacent}}$

COSECANT

csc $= \dfrac{\text{hypotenuse}}{\text{side opposite}}$

SECANT

sec $= \dfrac{\text{hypotenuse}}{\text{side adjacent}}$

COTANGENT

cot $= \dfrac{\text{side adjacent}}{\text{side opposite}}$

These ratios and the table of trigonometric functions found in the appendix are used to find the sides and angles of a right triangle.

Note: When solving problems, draw a right triangle and label the parts (hypotenuse, opposite, A, etc.). Then put the values that are given in the problem on the triangle. This will give a better picture of what you are looking for and what trigonometric function will give it to you.

PRACTICAL PROBLEMS

Note: For problems 1–4, round the answer to the nearer whole degree.

1. Find the measure of angle **A**. __________

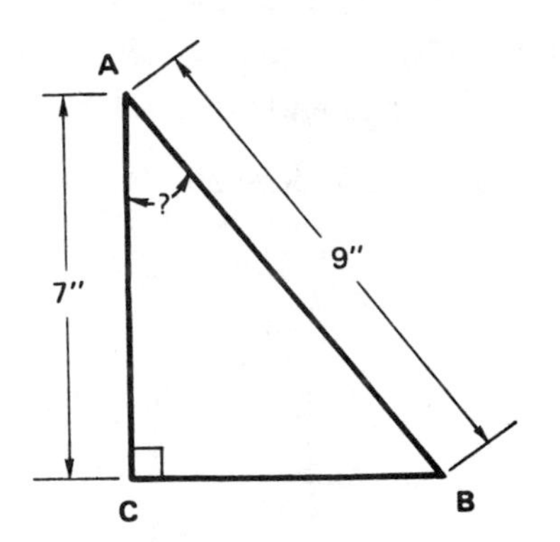

2. How many degrees are there in angle **E**? __________

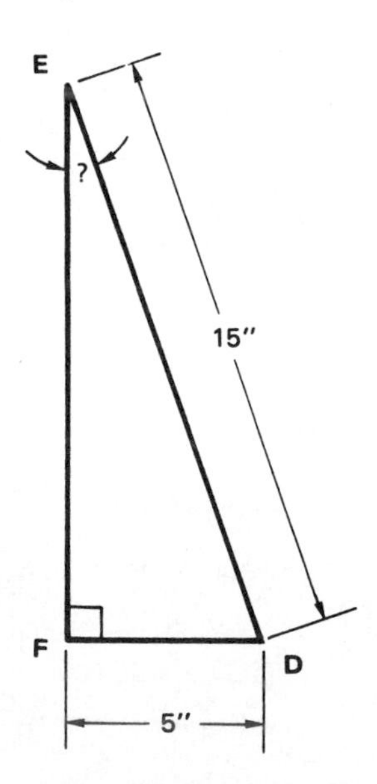

3. Determine the value of angle **Q**. __________

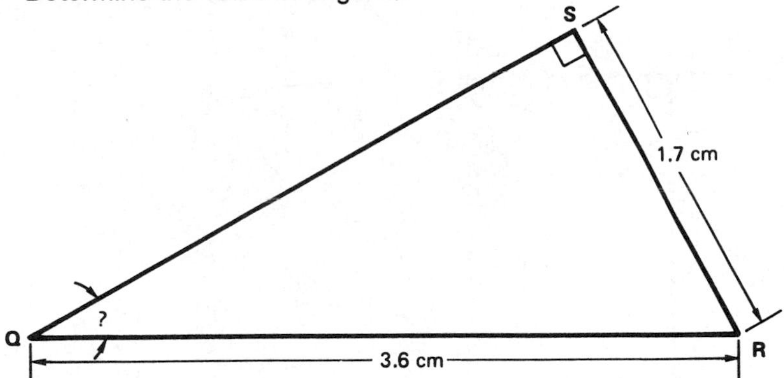

4. In triangle **XYZ**, side **XZ** is 1 1/2 feet long and side **YZ** is 2 1/4 feet long.

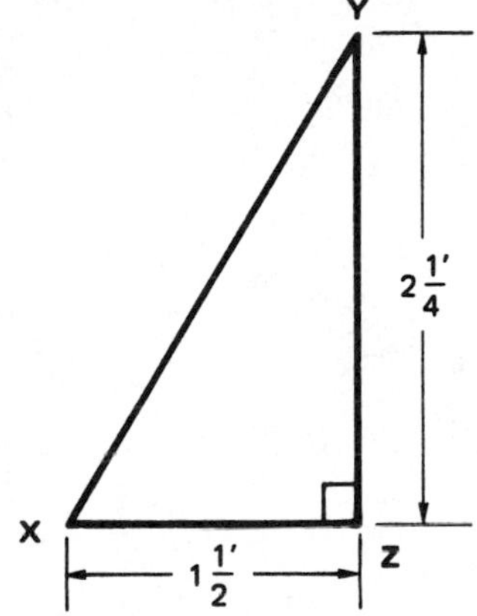

a. Find the measure of angle **X**. a. __________

b. Find the measure of angle **Y**. b. __________

Note: For problems 5–8, round the answer to the nearer hundredth.

5. How many centimeters long is side **AC**? __________

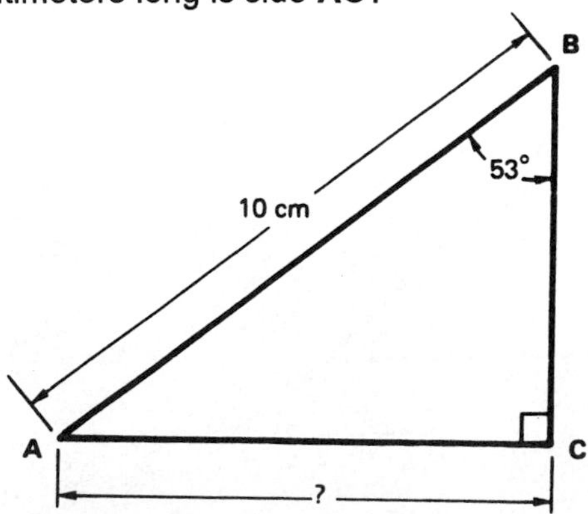

6. Find the length of side **EF**. ______

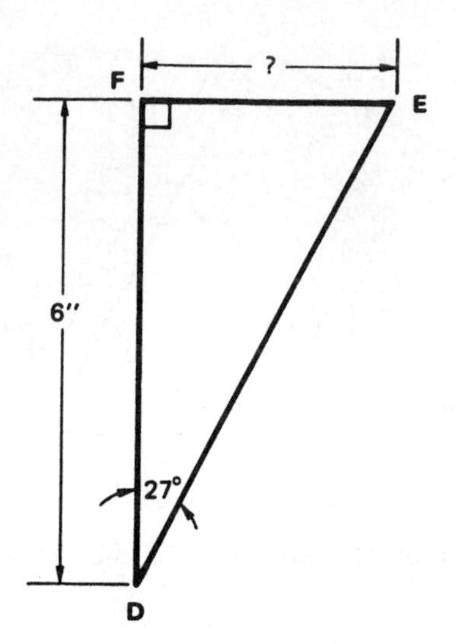

7. What is the length of the hypotenuse in triangle **MNO**? ______

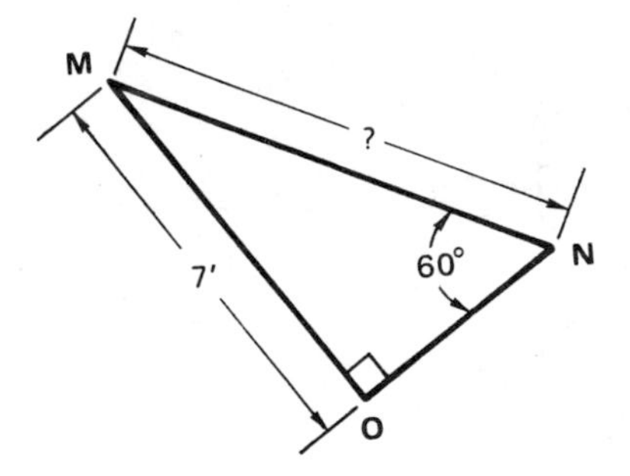

8. Find the length of side **TS**. ______

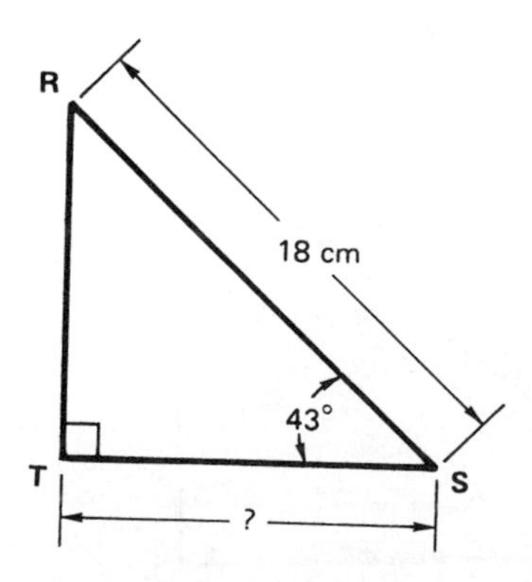

9. A fuel line for an oil burner is installed in a straight line from the tank to the burner. What is the vertical drop in the fuel line? Round the answer to the nearer hundredth foot. __________

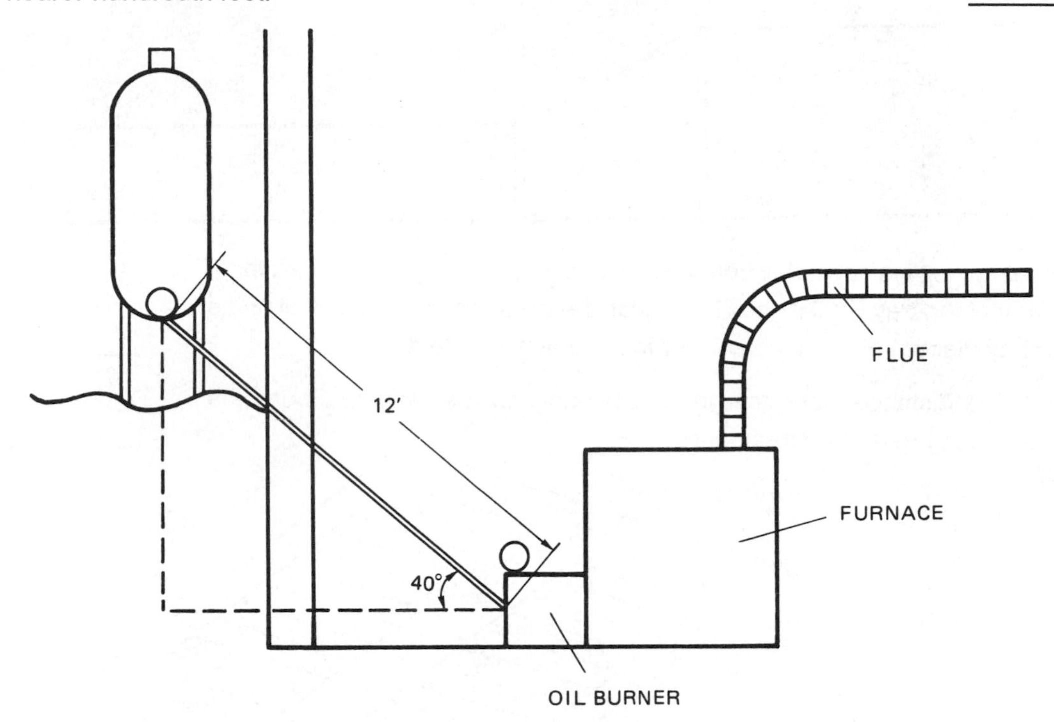

10. A round trunk duct fitting has a branch that forms an angle of 30° with the main duct. The main duct is straight. The main duct and the branch form a Y. How far apart are the centers of the ducts 25 feet from the Y along the center of the branch? Round the answer to the nearer hundredth foot. __________

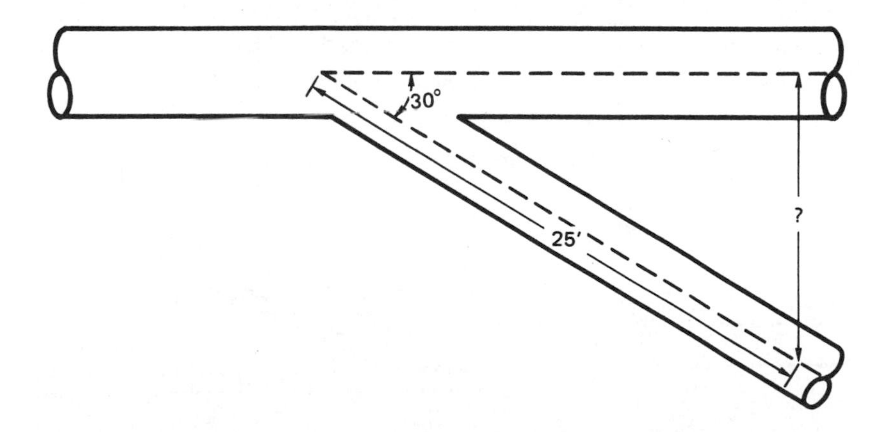

11. A section of duct is used to reduce the height of the duct. Find, to the nearer degree, the angle formed by the side of this reducer. ____________

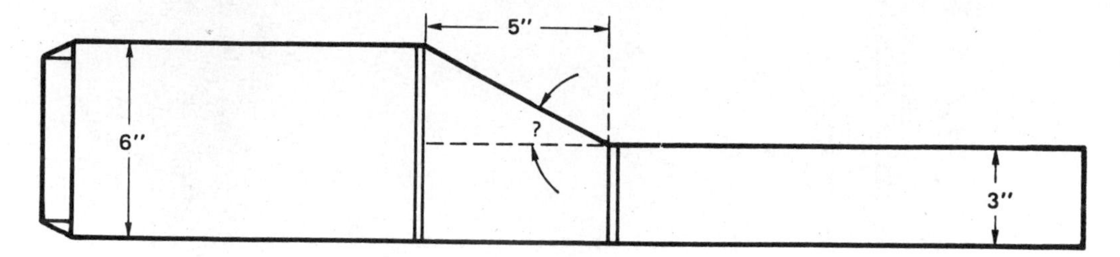

12. In an ice maker, ice cubes fall from a tray into a bin. The tray is 25 inches long. The front of the tray is 10 inches lower than the back of the tray. At what angle is the tray placed? Round the answer to the nearer whole degree. ____________

13. What is the distance along the grill of this baseboard spreader? Round the answer to the nearer hundredth inch. ____________

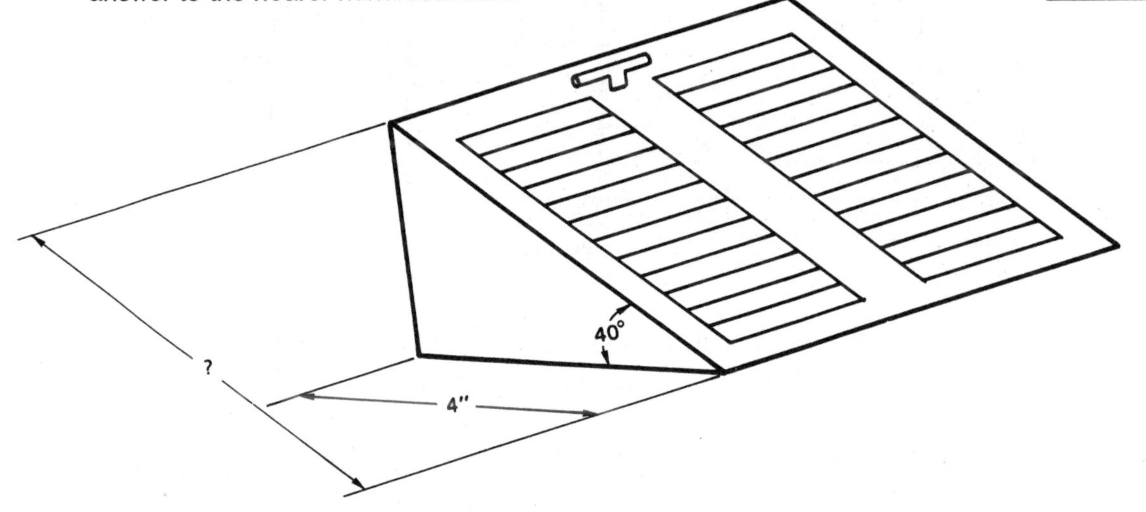

14. A compressor reed valve is made from a flat piece of metal. The reed is 0.8 centimeter long. When gas passes through the port, the movable end of the reed makes an angle of 25° with the normal position. How high is the movable end of the reed when the valve is open? Round the answer to the nearer hundredth centimeter. ____________

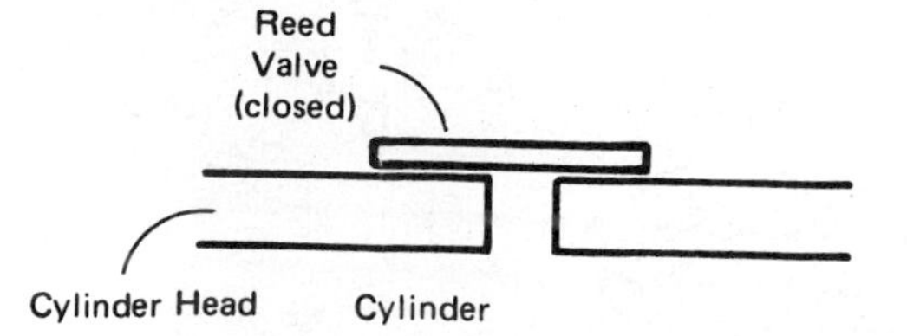

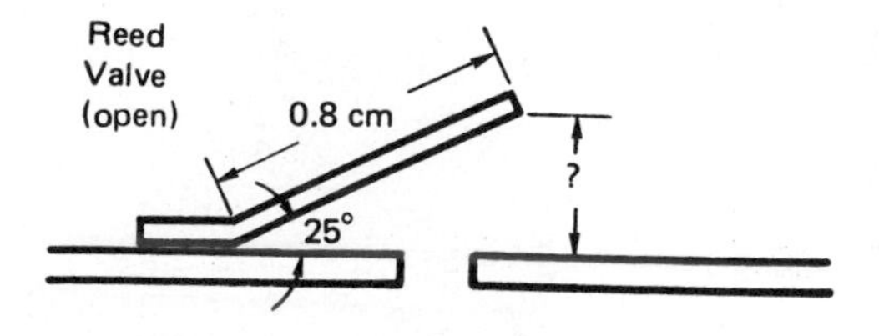

15. Find, to the nearer hundredth, the length of the side of the reducer of this duct. ______________

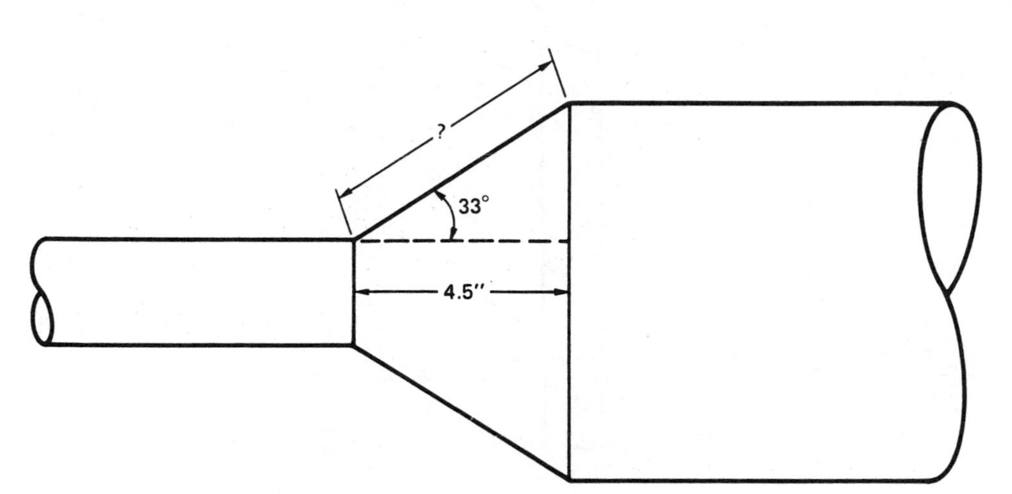

16. A support is built for a duct. Find, to the nearer degree, the angle that the support makes the duct. ______________

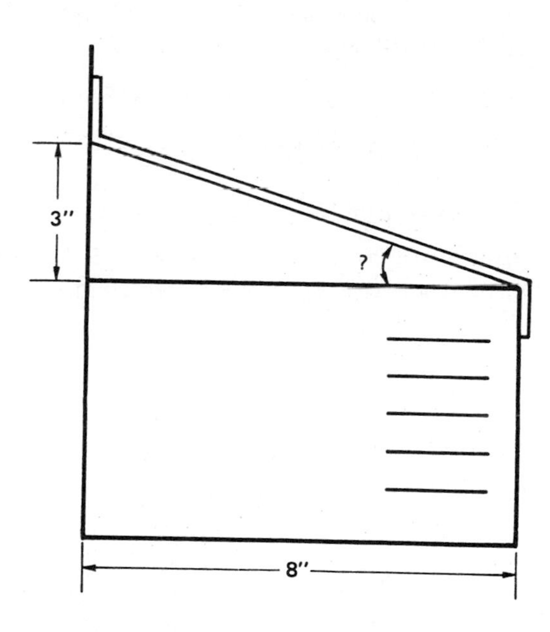

17. A branch duct passes through an opening. Find, to the nearer whole degree, the angle at which the branch leaves the main duct. __________

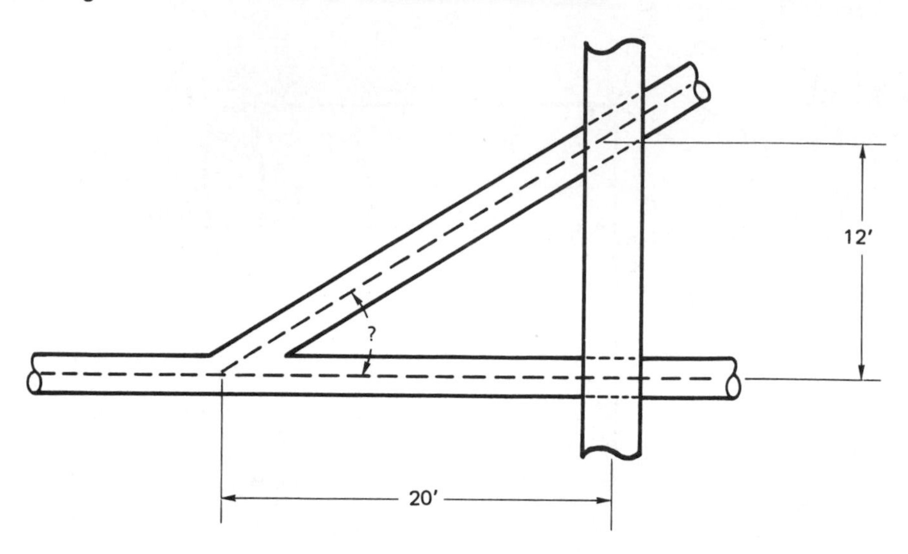

18. An oil burner nozzle is labeled 60°. This is the angle from straight ahead to the direction that the oil is sprayed. A piece of paper is placed 12 inches from the nozzle. How far away from the straight ahead mark does the oil strike the paper? Round the answer to the nearer hundredth inch. __________

19. A flue pipe joins the opening of an oil burner to the opening of a chimney. The chimney is 12 feet away from the oil burner. The opening on the chimney is 3 feet higher than the opening on the burner.
    a. What angle does the bottom of the flue pipe make with the horizontal? Round the answer to the nearer whole degree. a. __________
    b. Find, to the nearer hundredth foot, the length of the flue pipe between the chimney and the burner. b. __________

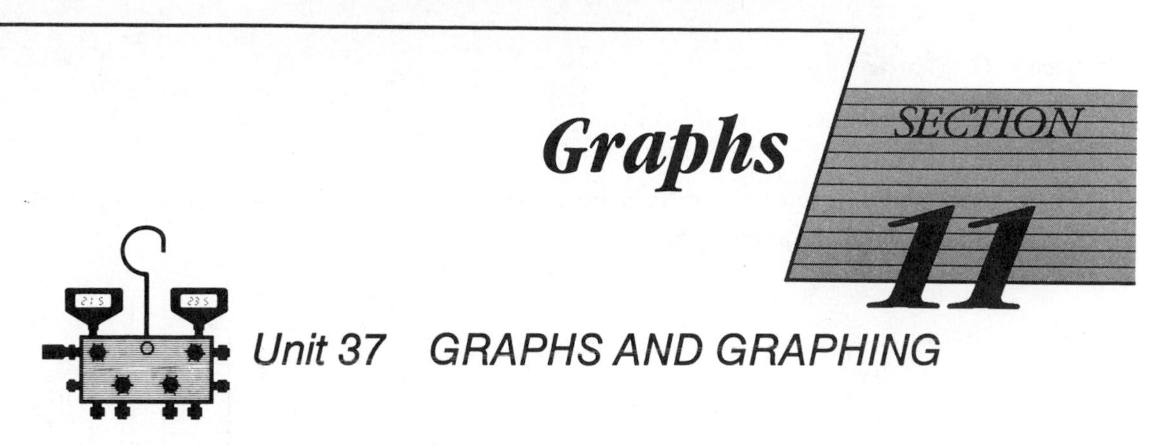

Graphs

SECTION 11

Unit 37 GRAPHS AND GRAPHING

BASIC PRINCIPLES

- Review and apply the principles of graphing to the problems in this unit.

Graphs are ways of easily displaying quantities of information. A single graph can display the relationship between two variables. Those two variables are given on the two axes of the graph. Each axis should be a uniform scale, but it is important to realize that the two axes do not have to have the same scale. Each axis should have the units and the scale displayed on it.

Additional information can be displayed on a single graph, but additional axes are needed, or additional lines (curves) must be put on the graph. In this case, care must be taken to follow the correct line to the correct axis to obtain the value that you are trying to find.

When drawing a graph, you must decide what you want to display. Then you have to determine the extent of your values. The whole graph should be filled with the values that are being displayed, not just a small portion of the graph.

Note: A graph may have many variables displayed on it at the same time. When reading such a graph, be certain that you are following the proper line to the correct scale to determine what the value is. A ruler or straight edge may often help when reading the graph, but be very careful using a straight edge when reading a chart that has curved lines.

The scale should tell you what the units of the graph are. Make sure that you include all of that information when you are drawing a graph. You also want to make sure that your data fills the graph, not just a very tiny portion of it, making the reading of your graph difficult.

PRACTICAL PROBLEMS

The graph below shows the number of calls made by the Warm and Cozy Heating Co. for a year. Questions 1–3 deal with the information on this graph.

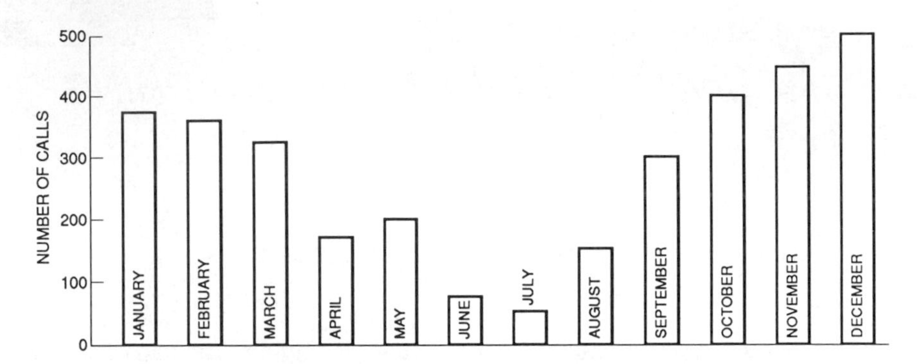

1. When would be the best time to overhaul the service trucks? ____________

2. An average repairman can make 3 calls in an 8-hour day.
   a. How many calls would a repairman make in a month? (An average work month has 22 days in it.) a. ____________
   b. What is the largest number of trucks (and repairmen) that are needed by the Warm and Cozy Co.? b. ____________

3. It has been found that during June and July running a "Heating System Tune-up Special" will increase the number of calls made by 250%. If that special were run this year, how many repairmen would be needed during the months of June and July—that is, how many repairmen needed in problem 2 could be allowed to take vacation together? ____________

Questions 4–5 refer to the graph for the Keep Kool Co.

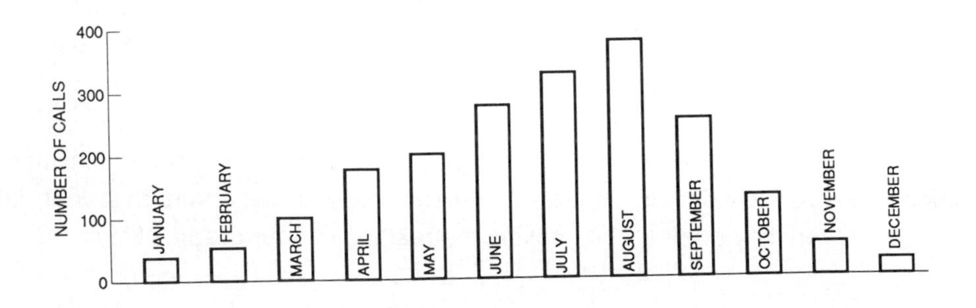

4. How many repair calls were made in the month of September? ________

5. If an average repairman makes 3 calls in an 8-hour day, what is the greatest number of repairmen that are needed and what is the smallest number? ________

It is decided to merge the Warm and Cozy Heating Co. and the Keep Kool Co. to form the Hot and Cold Service Co. Questions 6–7 refer to the graphs that were used in the previous questions.

Question 6–9 deal with same information.

6. Make a bar graph of the combined Number of Service Calls on a monthly basis for the Hot and Cold Service Co. ________

7. Make a line graph of the same information in problem 6. ________

8. If part-time repairmen can be hired, what is the smallest number of full-time repairmen needed? ________

9. What is the largest number of part-time repairmen needed? ________

Questions 10–13 deal with a simplified pressure heat diagram for R-12 refrigerant.

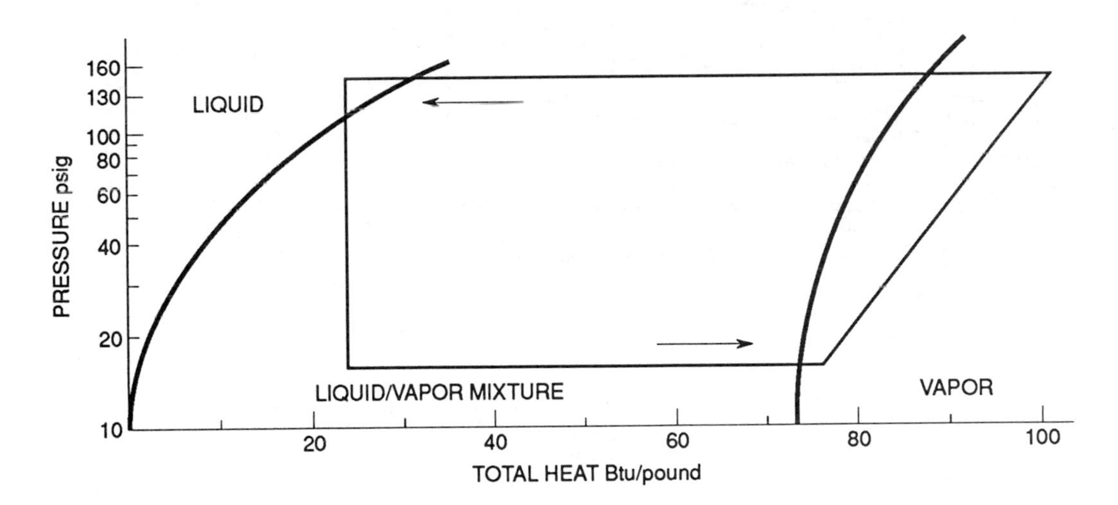

10. A refrigerator uses 8 ounces in its system. How much heat is given up by the R-12 when all of it is cooled from a compressed vapor to a liquid? ________

11. How much heat does the full 8 ounces remove from the refrigerator area as it boils back to a vapor? ________

12. How many times must the 8-ounce charge of refrigerant go through the cycle to remove the heat put into the refrigerator when hot gelatin was put into the refrigerator to cool? The 4 pounds of liquid went from 180°F (720 Btu) to 60°F (240 Btu). Round off to the nearer tenth. __________

13. How much heat is added by the pump to 1 pound of the refrigerant and must be removed without doing any cooling? __________

Questions 14–16 deal with the charging chart for a heat pump system below.

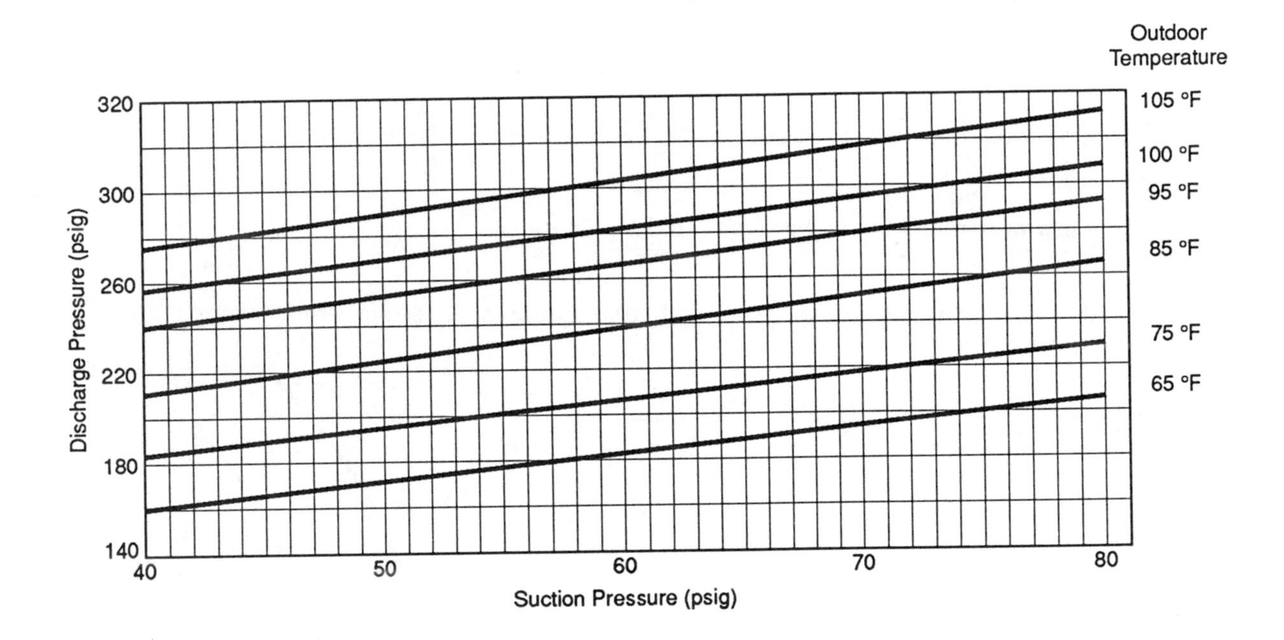

14. If the compressor suction pressure is 70 pounds per square inch gauge pressure (psig) and the outside temperature is 85°F, what should the discharge pressure be reading? __________

15. If a system is overcharged, the discharge pressure will read higher than it should. If the system is undercharged, the discharge pressure will read lower than it should.
    The compressor suction pressure is 62 psig and the discharge pressure indicates 290 psig when the outside temperature is 85°F. Should the refrigerant be added to the system or taken out? __________

16. A correctly charged system is checked in the morning when the outdoor temperature is 75°F. The suction pressure indicates 58 psig. A second check of the system is made on an afternoon when the outdoor temperature is 90°F. There was no change in the suction pressure reading. What should the old and new discharge pressures be? ____________

Questions 17–18 deal with the abbreviated psychometric chart below.

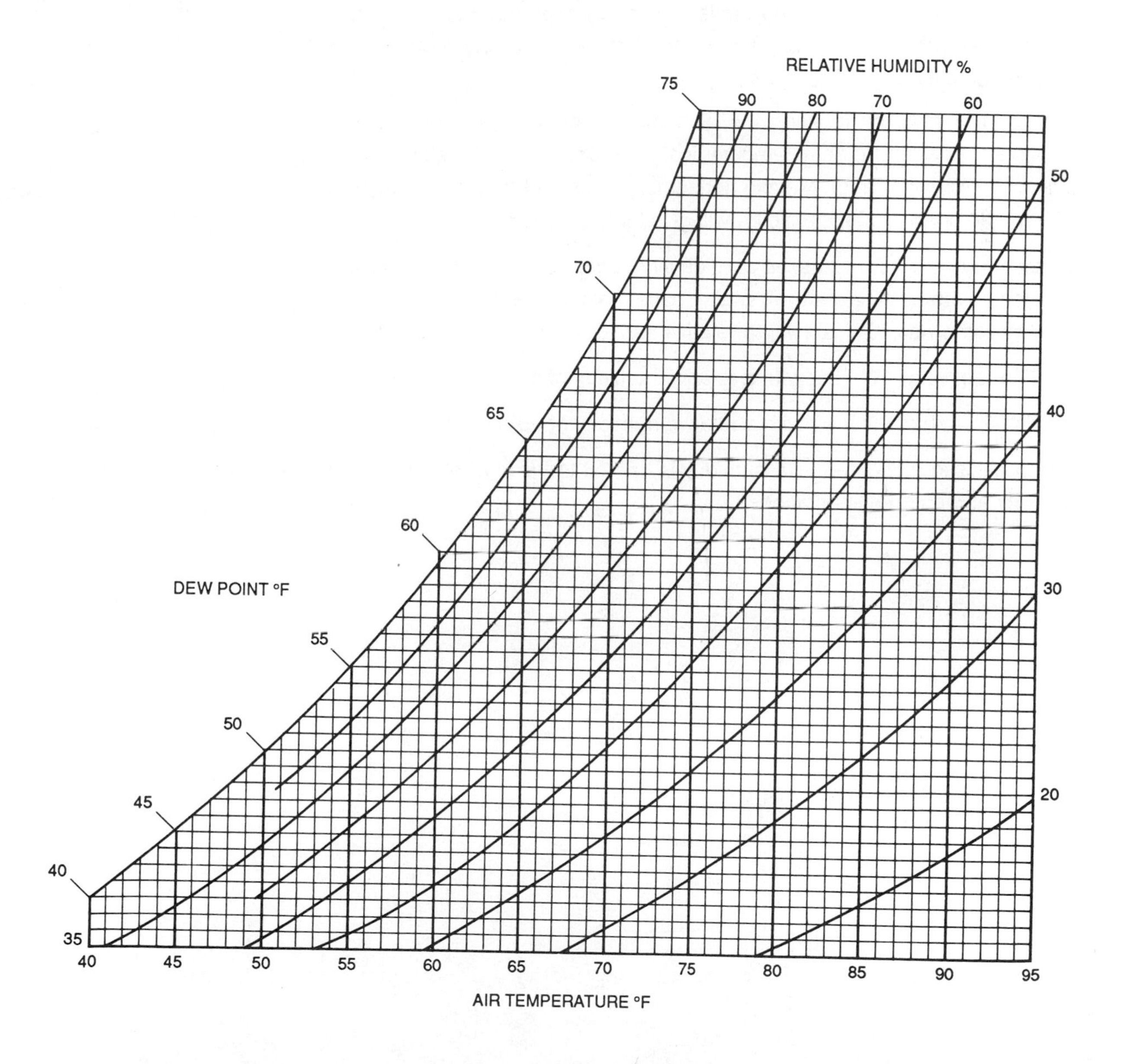

The dew point is the temperature of a quantity of air where it cannot hold any more water vapor. Relative humidity is the amount of water vapor in the air compared to the maximum amount the air could hold at that temperature. Heating the air up without adding moisture (water vapor) will lower the relative humidity; cooling the air down will raise the relative humidity.

If air is cooled below its dew point, moisture will come out of the air. This is what happens on glasses of cold drinks or cold pipes in the summer. It can also happen to air-conditioning ducts running through warmer rooms or air spaces. The psychometric chart gives information about the dew point and relative humidity.

The chart is read by locating the air temperature along the bottom scale (horizontal axis) and then moving vertically upward until the relative humidity is reached (curved lines). Moving horizontally to the left from that point to the edge of the chart (100% humidity) will give the dew point.

17. A duct carrying 55°F air passes through air spaces listed below. In each case determine whether the duct will sweat (have moisture form on it).
    a. The room has 70°F air with a relative humidity of 50%. a. ____________
    b. The room has 85°F air with a relative humidity of 50%. b. ____________
    c. The room has 90°F air with a relative humidity of 35%. c. ____________

18. If 65°F air with a 60% relative humidity is heated to 85°F, what is its relative humidity at these new conditions? ____________

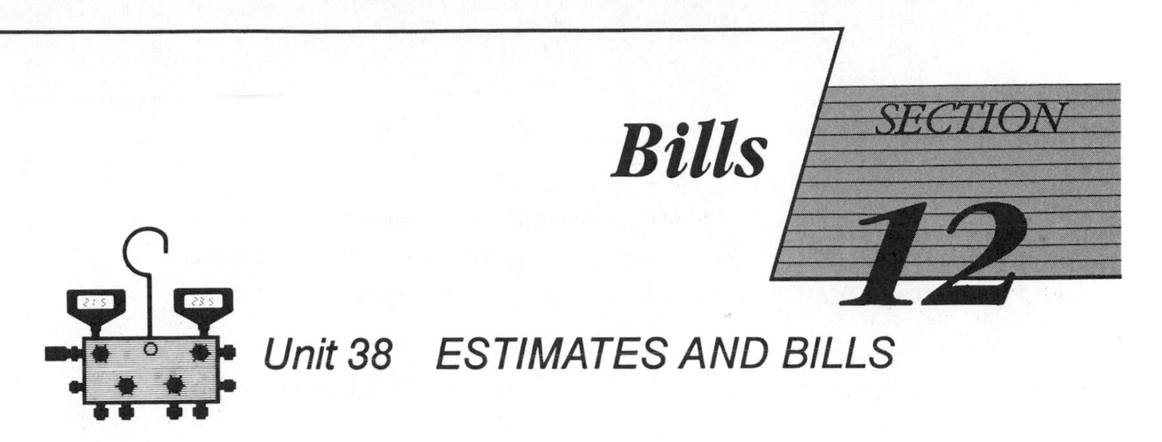

Bills

SECTION 12

Unit 38 ESTIMATES AND BILLS

BASIC PRINCIPLES

- Study and apply the principles of estimates and bills to the problems.

Customers often ask for itemized bills after work has been completed. At times they also want a written estimate for a possible job. It is important to be able to give these to the customer. The problems below are designed to give practice in writing estimates and bills. For each problem fill in the form and answer the related questions.

A bill is filled out for a number of reasons. One is to give you a record of what inventory was used. A bill also gives a record of how much time workers spent on various jobs. It also can be used by the customer to see what pieces were replaced.

To make it clear to everyone using the bill as a record, each bill should be clearly filled out. Make sure you write a clear description of the pieces used, the labor involved, or the basic charge made. Make sure that you put down the price per item when multiple items are used. This should go in the price column. The price per item times the number of items is the amount charged and should go in the amount column.

If the bill is an estimate, be sure to clearly mark that on the bill. It is also a good idea to put down how long the customer has to decide if they want the job done for that price.

When filling out bills and estimates it is important to fill out as much as possible and to be as clear as you can. You want someone else to read the sheet and understand it completely. Take your time to be certain that there is nothing that is uncertain or unclear.

PRACTICAL PROBLEMS

1. A customer asks for an estimate for the installation of a central air conditioner. A 3-ton unit is required. The estimate will show the estimated costs for parts and labor separately. The word "estimate" is to be written clearly on the form.

   A 3-ton unit $1 469.00
   Labor (15) hours . . . . $11.00 per hour

Keep Kool Refrigeration Co.
1000 Main St.
Anytown, SC 29000

Customer:

Date
Salesperson
No. 3149

QUANTITY	ITEM	PRICE	AMOUNT
ORIGINAL	**Thank You**		

a. If a 4% tax on parts is added, what is the additional charge? a. ____________

b. Two installers will install the unit. Each will work the same amount of time and will be paid $8.00 per hour. How much will each installer earn from this job? b. ____________

2. A refrigerator is repaired using the following:

   1 compressor $153.75
   1½ pounds refrigerant R-12 . . . . $1.10 per pound
   1 thermostat control $13.95
   nuts, washers, grease $0.80
   5 hours labor . . . . $9.50 per hour

Keep Kool Refrigeration Co.
1000 Main St.
Anytown, SC 29000

Customer:

Date
Salesperson
No. 3149

QUANTITY	ITEM	PRICE	AMOUNT
ORIGINAL	Thank You		

a. If the refrigerator owner did the work without hiring a repairer, what would be the cost of the parts alone? a. ________

b. The repairer is paid $7.25 per hour. How much money does the company get for the labor on this job? b. ________

c. How much does the repairer make on this job? c. ________

3. The installation of a 3 1/2-ton central air conditioning system needed the following:

1 compressor, condenser, and evaporator	$942.00
1 concrete slab for condenser	$32.00
2½ pounds refrigerant R-11 charge	$1.20 per pound
2 ft drainpipe (1-inch diameter)	$0.42 per ft
30 ft connecting tubing	$0.79 per ft
1 plenum	$47.50
1 roll duct insulation	$52.95
1 roll duct tape	$2.49
1 thermostat	$28.95
35 ft thermostat wire	$5.95
Labor (3 installers—8 hours each)	$9.75 per hour

Keep Kool Refrigeration Co.
1000 Main St.
Anytown, SC 29000

Customer:

Date
Salesperson
No. 3149

QUANTITY	ITEM	PRICE	AMOUNT
ORIGINAL	**Thank You**		

a. It took one installer 7 hours to insulate the ducts using the insulation and tape. How much would be saved if the ducts were already insulated? a. ____________

b. The company offers a 6% discount if paid in cash. How much is saved by paying cash? b. ____________

4. A repairer makes an annual check on an oil burner. The charges for items and services were:

   1 gun nozzle $4.75
   1 air filter (14 in × 20 in) $1.45
   1 fuel filter cartridge $2.95
   Clean burner and make adjustments $9.00
   Check for water in fuel tank $4.25
   House call $15.00

   a. Four dollars of the house call charge went for truck expenses. The remainder of the house call charge and the other costs (except parts) are for labor. If the repairer took 1 1/2 hours to do the job, what is the labor cost per hour? a. ____________
   b. If the homeowner changes the filters beforehand, what would be the charge for the visit? b. ____________

Jones Heating Co.
100 Center St.
Yourtown, PA 18000

Customer:

Date
Salesperson
No. 4285

QUANTITY	ITEM	PRICE	AMOUNT
Customer's Copy	**Thank You**		

5. An oil burner needed repairs. The repairer replaces the parts listed:

1 gun nozzle	$3.45
25 ft of 3/8-in copper fuel line	$0.69 per ft
1 fuel filter	$6.95
1 new oil pump	$43.95
1 pair oil gun electrodes	$2.35 each
Labor (2 hours)	$7.25 per hour

Jones Heating Co.
100 Center St.
Yourtown, PA 18000

Customer:

Date
Salesperson
No. 4285

QUANTITY	ITEM	PRICE	AMOUNT
Customer's Copy	**Thank You**		

a. A rebuilt oil pump costs $24.00. What is the bill if a rebuilt pump is used? a. ____________

b. A 4% sales tax is charged for parts. How much does this add to the original bill? b. ____________

23. A capillary tube has an inside diameter of 0.031 inch and an outside diameter of 0.083 inch. Find the wall thickness of the tube. ________

(16)

24. An outside thermometer reads 33°C. What is this temperature expressed in degrees Fahrenheit? ________

(25)

25. The standard evaporating temperature of a refrigeration cycle is 5°F. Find the equivalent standard evaporating temperature on the Celsius scale. ________

(25)

26. Three electric heaters in a forced air system are wired in a series circuit. The heaters have resistances of 2.09 ohms, 5.57 ohms, and 6.96 ohms. What is the total resistance of the circuit? ________

(11, 30)

27. A light bulb in a refrigerator uses 120 volts and draws a current of 1/3 ampere. The door switch fails and the light bulb remains on when the door is closed.
    a. Find, in watts, the power that the light bulb uses. a. ________
    b. Each hour, 1 watt of power produces 3.415 Btu of heat. This extra heat must be removed. How much heat must be removed each hour? b. ________

(13, 30)

28. Refrigerant gas enters the compressor at 14.7 psia. The gas leaves the compressor at 111.72 psia. The volume of the gas that leaves the compressor is 0.9 cubic inch. The temperature remains the same. Find, in cubic inches, the volume of the gas that enters the compressor. ________

(31)

29. A gauge on the high pressure side of a refrigerator reads 28.753 psig. The atmospheric pressure is 14.7 psi. What is the absolute pressure? ________

(31)

30. The bill for repairing a refrigerator is $72.35. The state sales tax is 4%. What is the total amount of the bill? Round the answer to the nearer whole cent. ________

(17, 38)

31. To check the efficiency of an oil burner, 12 cubic centimeters of flue gas are drawn. After the carbon dioxide is removed, 10.86 cubic centimeters of flue gas remain. What percent of the flue gas is carbon dioxide? ________

(17)

32. The tank of an oil burner contains 250 gallons of oil. The nozzle on the gun reads 1.2 gallons per hour. Find the number of hours the burner can run on one tank of oil.
(14) ____________

33. A compressor takes 7.348 6 cubic inches of refrigerant and compresses it to 0.812 cubic inch. The ratio is the compression ratio. The second term of the compression ratio is 1. Find the compression ratio of this compressor.
(19) ____________

34. When full, a storage cylinder for refrigerant holds 150 pounds of refrigerant. There are 60 pounds of refrigerant left in the cylinder. What is the ratio of the refrigerant left to the refrigerant used?
(19) ____________

35. The scale on a drawing is 1/4 inch equals 1 foot. On the drawing, a duct is 4 5/8 inches long. Find the actual length of the duct.
(20) ____________

36. A strap for sealing a special duct joint is needed. There is a 3/4-inch lap on the strap. How long is the strap?
(33)

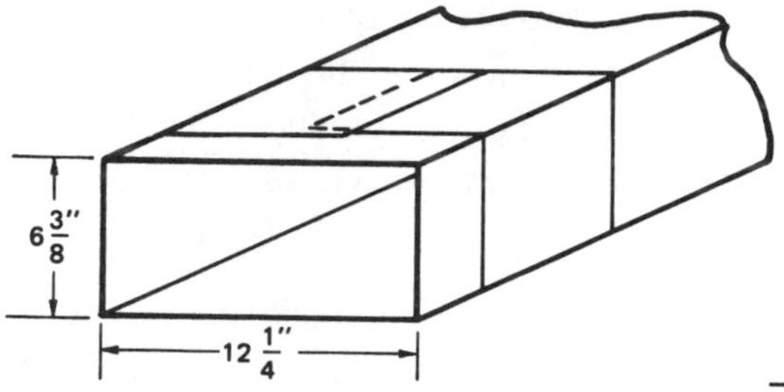

$6\frac{3}{8}''$

$12\frac{1}{4}''$

____

37. A 15-foot roll of insulation has a 76-inch long piece cut from it. What is the length of the insulation remaining?
(21, 23) ____________

38. To seal the duct joint on this duct, tape is wrapped around the duct. No lapping is needed. Find, to the nearer thousandth meter, the length of tape to be used.
(21, 23) ____________

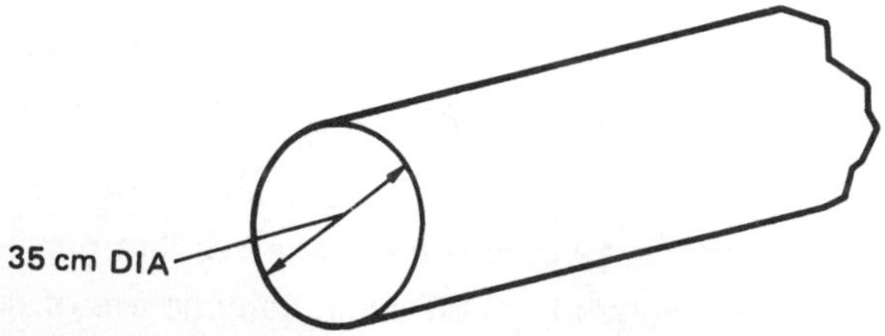

35 cm DIA

39. An installer has an 8-inch diameter duct and a 20-centimeter diameter duct. Which duct has the larger diameter?
(22) ____________

40. For proper airflow, the area of the opening of a rectangular duct must be 143 3/8 square inches. The height of the duct must be 7 3/4 inches. Find the width of the duct.

(26)

41. Find, in meters, the area of this floor.

(26)

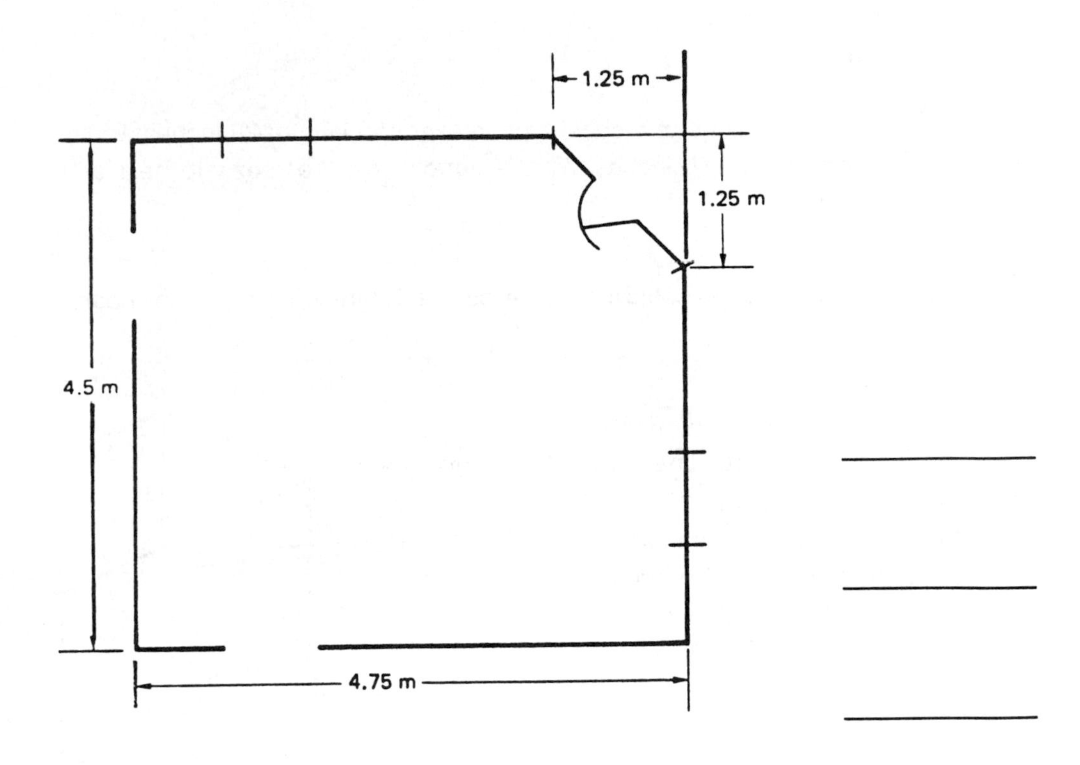

42. A circular duct has a diameter of 8 5/8 inches. Find the area of the duct opening to the nearer thousandth square inch.

(26)

43. The major heat loss in a room is a window. The window is 1.8 meters high and 75 centimeters wide. What is the area of the window in square meters? __________
(21, 26)

44. Find the volume of this liquid receiver. Round the answer to the nearer hundredth cubic inch. __________
(29)

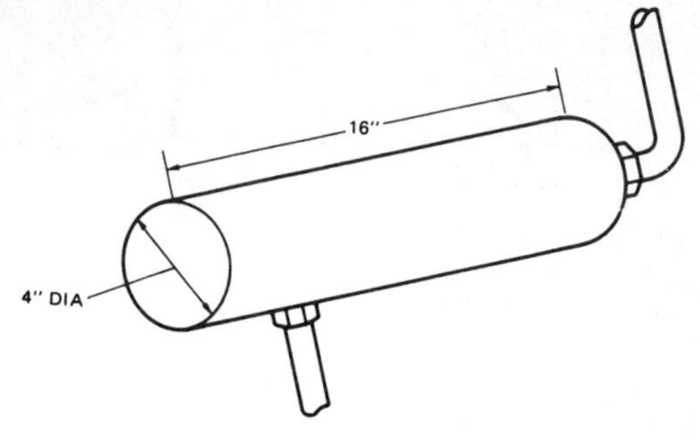

45. The inside dimensions of a chest-type freezer are: length, 29 inches; width, 25 inches; height, 36 1/2 inches. What is the volume of the freezer to the nearer tenth cubic foot? __________
(28)

46. A fan is rated at 220.74 cubic meters per minute. Express this rate to the nearer cubic foot per minute. __________
(28)

47. Air enters this room through an 8-inch circular duct. The rate of airflow is 250 feet per minute. How long does it take to change the air in the room? Round the answer to the nearer hundredth minute.
(28, 29)

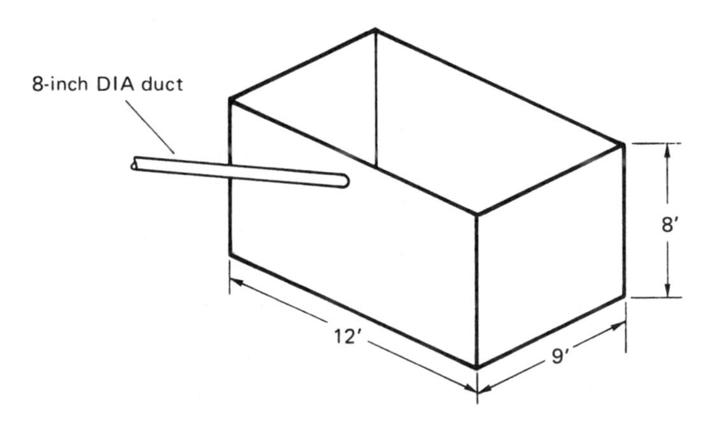

____

48. A 5-foot long duct is to be made. The duct is to be 6 inches high and 14 inches wide. A grooved seam is to be used. The seam allowance is 3/8 inch.
    a. Find the length of material needed to make this duct. a. __________
    b. Find the width of material needed to make this duct. b. __________
(33)

49. What is the arc length of the center of this duct? Round the answer to the nearer hundredth foot.

(34)

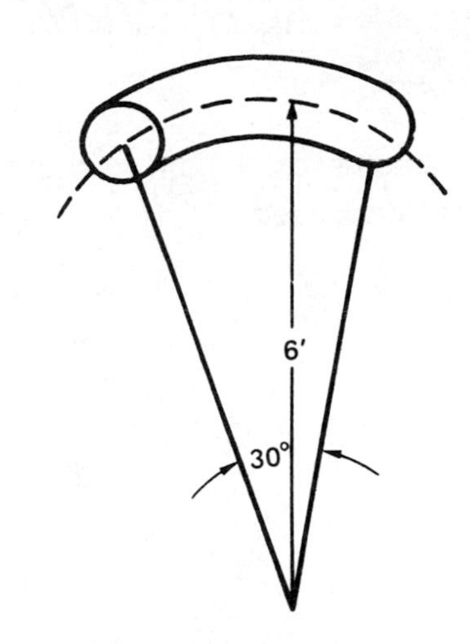

____

50. This elbow is 8 inches wide. The radius of the throat is 5 inches.
    a. What is the arc length of the throat to the nearer hundredth inch?
    b. What is the arc length of the heel to the nearer hundredth inch?

(35)

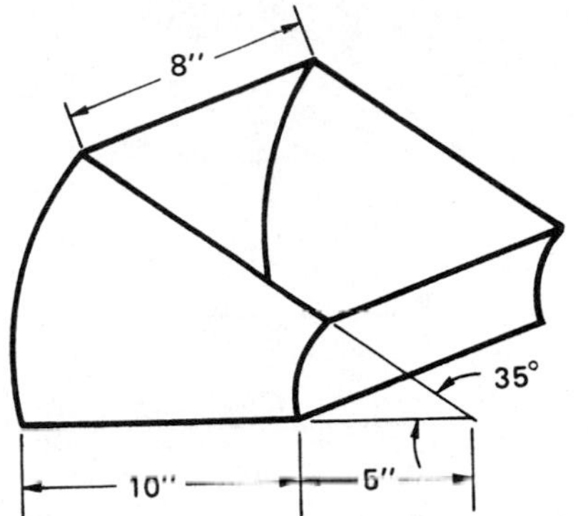

a. __________

b. __________

51. Find the angle through which this duct is bent. Round the answer to the nearer whole degree.

(36)

____

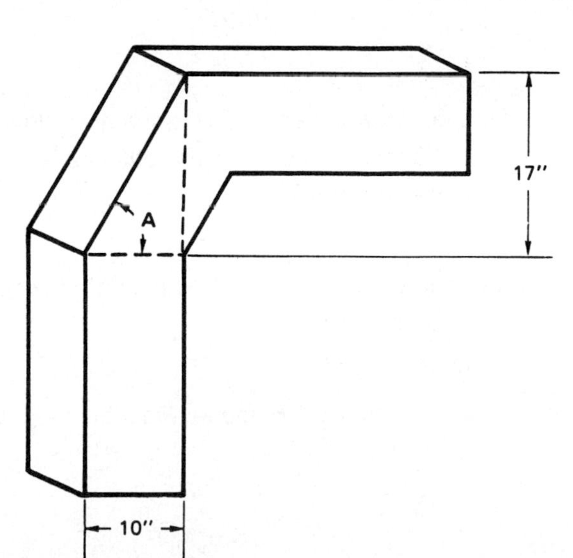

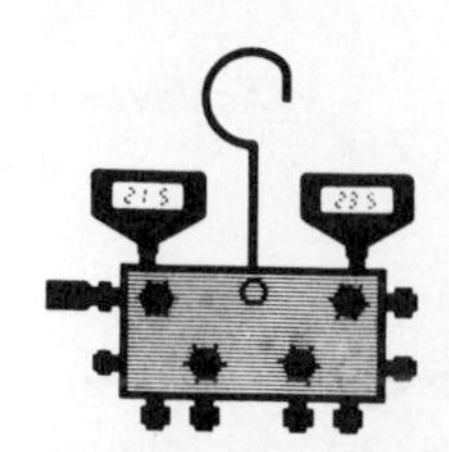

ACHIEVEMENT REVIEW B

Note: The numbers in parentheses, ( ), given below each question show the unit in which similar problems have been discussed.

1. (1)
$$\begin{array}{r} 2\,790 \\ +\,4\,253 \\ \hline \end{array}$$

2. (2)
$$\begin{array}{r} 6\,047 \\ -\,1\,158 \\ \hline \end{array}$$

3. (3)
$$\begin{array}{r} 726 \\ \times\ \ 78 \\ \hline \end{array}$$

4. (11)
$$\begin{array}{r} 27.409 \\ 4.386 \\ +\ \ 0.041\,3 \\ \hline \end{array}$$

5. (12)
$$\begin{array}{r} 193.46 \\ -\ \ 64.877 \\ \hline \end{array}$$

6. (13)
$$\begin{array}{r} 6.28 \\ \times\,4.13 \\ \hline \end{array}$$

7. $35\overline{)24\,815}$
(4)

8. $6.27\overline{)25.268\,1}$
(14)

9. 5/6 + 3/4 ____________
(6)

10. 1 2/3 + 2 1/2 ____________
(6)

11. 9/10 – 7/8 ____________
(7)

12. 3 1/5 – 1 5/6 ____________
(7)

13. 2/3 × 9/14 ____________
(8)

14. 2 2/7 × 2 11/12 ____________
(8)

15. 1/3 ÷ 2/5 ____________
(9)

16. 4 1/5 ÷ 1 1/3 ____________
(9)

17. A window air conditioner is 22 3/4 inches wide. When installing it in a window, two fillers are needed. One filler is 6 7/8 inches wide and the other is 6 1/2 inches wide. What is the width of the window opening? ____________
(6)

18. A tank contains 20 pounds of refrigerant. Charges of 1 2/3 pounds are taken from the tank. How many charges can be made? ____________
(9)

19. The weight of one foot of 3/8-inch copper tubing is 1/5 pound. Find the weight of 42 1/3 feet of tubing. ____________
(8)

20. A 6 1/2-inch duct must pass through a wall. A 4 7/8-inch hole is already in the wall. How much must the diameter of the hole be enlarged? __________

(7)

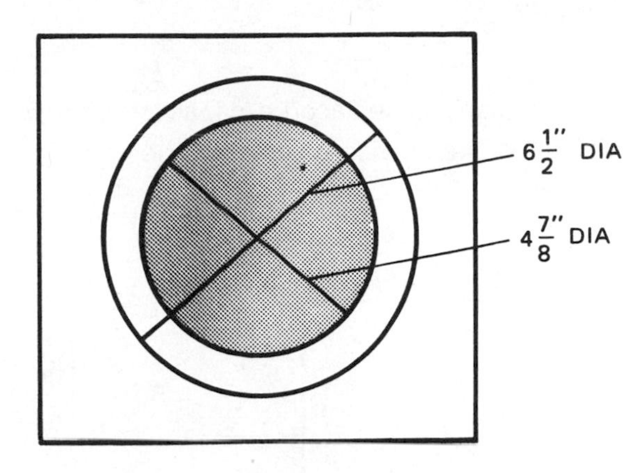

21. The total resistance of a substance is the sum of the *R* values for each material. The materials and *R* values for a certain wall are: brick veneer, 0.56; concrete block, 2.5; 3/4-inch polystyrene insulation, 3.125; inside wallboard, 0.62. What is the total resistance of the wall? __________

(11)

22. Pieces of 4.7-inch long duct tape are needed for an installation job. How many 4.7-inch pieces can be cut from a 53.4-inch piece of tape? __________

(14)

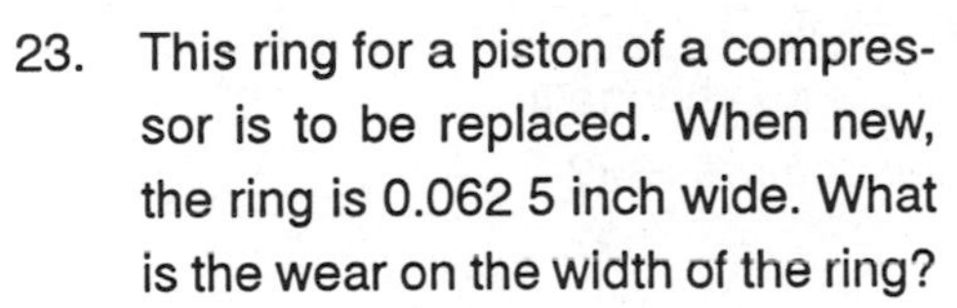

23. This ring for a piston of a compressor is to be replaced. When new, the ring is 0.062 5 inch wide. What is the wear on the width of the ring? __________

(12)

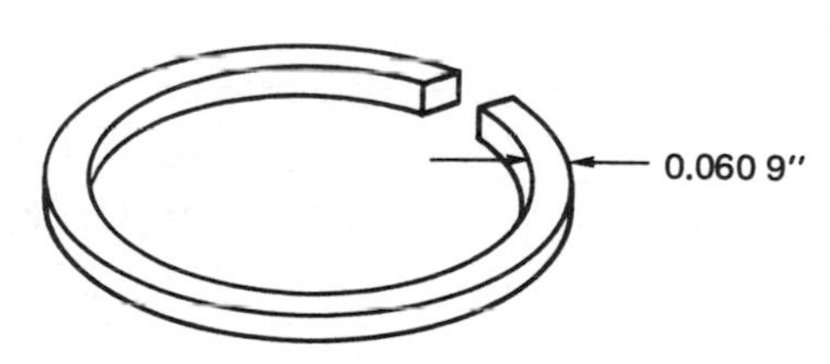

24. Find the outside diameter of this tube. Express the answer as a decimal fraction. __________

(15, 16)

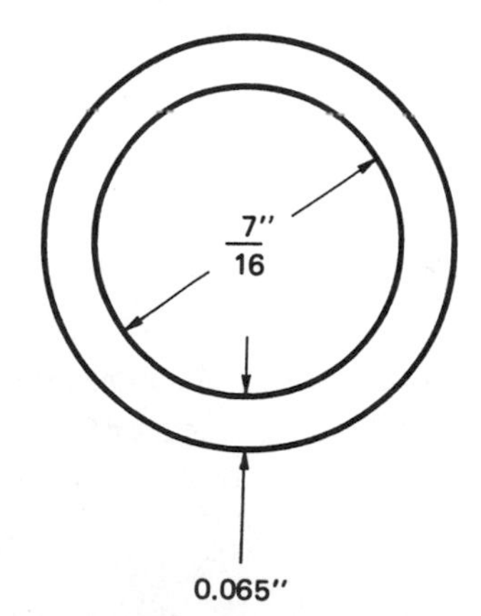

25. The circulation fan in a forced air system turns on automatically. It turns on when the temperature in the air jacket reaches 160°F. What is this temperature to the nearer tenth degrees Celsius? __________

(25)

26. A motor is being serviced. The last servicer replaced a capacitor with two capacitors wired in parallel. This time, the servicer uses only one capacitor. What size capacitor should be used? __________

(30)

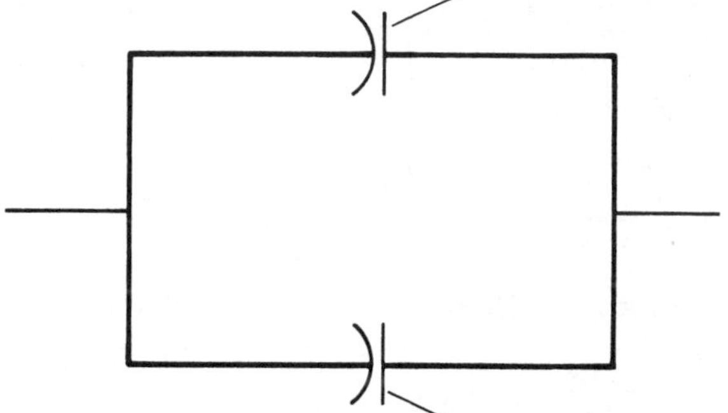

60 MICROFARADS

24 MICROFARADS

27. A refrigerator with automatic defrost has an electric heater that defrosts the evaporator coils. The heater is rated at 120 volts and draws 1.85 amperes of current. What power is used when this heater is on? __________

(30)

28. A compressor takes refrigerant at 12.8 psia and compresses it. The gas is compressed from 1.2 cubic inches to 0.15 cubic inch. There is no change in temperature. What is the pressure of the gas as it leaves the compressor? __________

(31)

29. The values for the low pressure and high pressure sides of an air conditioning compressor are given. Find the missing value. __________

LOW PRESSURE SIDE	HIGH PRESSURE SIDE
P = 12 psia	P = ?
V = 1.5 cu in	V = 0.6 cu in
T = 27°C	T = 57°C

(31)

30. Refrigerant leaves a compressor with a gauge pressure of 122.45 pounds per square inch. What is the absolute pressure of the gas? __________

(31)

31. A thermostatic expansion valve is needed to repair a freezer. The valve can be purchased from a dealer who gives an 8% discount. The price of the valve is $34.74. What is the discounted price? __________

(18)

32. The installation of an air conditioning system in a building will take 560 hours. Installers have spent 476 hours on this job. What percent of the job is completed? __________

(17)

33. Refrigerant R-503 is a mixture of refrigerants R-13 and R-23. Each pound of refrigerant R-503 contains 59.9% of refrigerant R-13 and 40.1% of refrigerant R-23.
    a. How many pounds of refrigerant R-13 are there in 5 pounds of refrigerant R-503? a. __________
    b. How many pounds of refrigerant R-23 are there in 5 pounds of refrigerant R-503? b. __________

(17)

34. A capacitor of a motor can be used if it has a value near the printed value. An acceptable value is not more than 10% below the printed value nor 20% higher than the printed value. A printed value is 125 microfarads.
    a. What is the lowest acceptable value? a. __________
    b. What is the highest acceptable value? b. __________

(17)

35. An oil furnace is located in a basement. The ceiling of a basement is 7 feet 10 inches high. The furnace stands 6 inches off the floor and is 4 feet 8 1/2 inches high. What is the clearance between the top of the furnace and the ceiling? __________

(10)

36. A piece of copper tubing, 2 feet 4 1/2 inches long, is needed to repair a refrigerator. The tubing is cut from a 25-foot coil. How much tubing is left? __________

(21, 23)

37. To make a duct, 24 pieces of ducting are used. The fitted length of one piece is 84.95 centimeters. All of the other pieces have a fitted length of 132.7 centimeters each. Find, in meters, the length of the duct. __________

(23)

38. A circular duct has a circumference of 14.13 centimeters. Find the diameter of the duct. Round the answer to the nearer hundredth centimeter. __________

(23)

39. A piece of 1.75-meter long sheet metal is used to make 4-inch by 8-inch rectangular ducts. Each duct is to be 5 feet long. Find, in feet, the amount of material that must be cut from the length. __________

(22, 23)

40. Find, in square feet, the area of this floor plan. __________

(26)

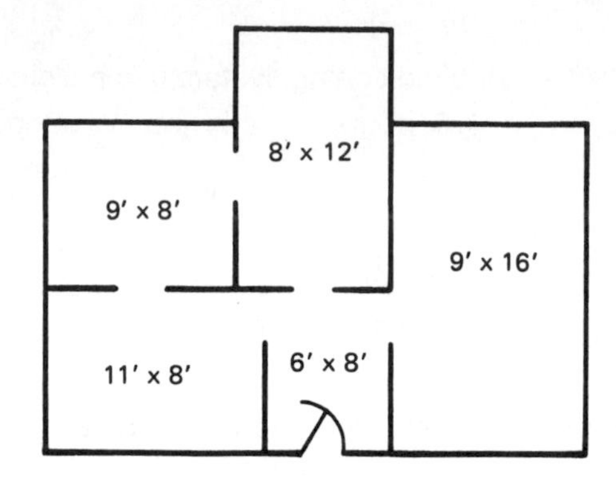

41. A special triangular duct is made. What is the area of the opening in the duct? __________

(26)

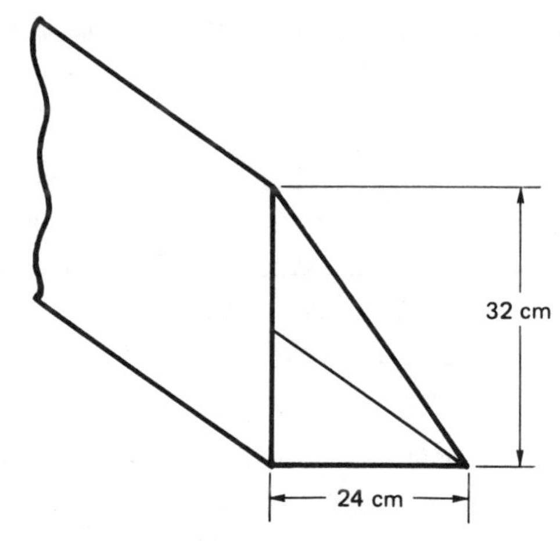

42. A piston has a diameter of 18 centimeters. Find, to the nearer thousandth square centimeter, the area of the top of the piston. __________

(26)

43. The cylinder of a compressor is 0.84 inch in diameter. When the piston is in its lowest position, the height of the opening is 0.9 inch. What is the volume of the cylinder? Round the answer to the nearer hundredth cubic inch. __________

(29)

44. A compressor has a listed displacement of 25 cubic centimeters. What is the displacement of the compressor in cubic inches? Round the answer to the nearer hundredth cubic inch. __________

(29)

45. A motor turns at 1 200 revolutions per minute. The wheel on the motor has a diameter of 1 3/4 inches. The fan turns at 525 revolutions per minute. What is the diameter of the wheel on the fan? __________

(20)

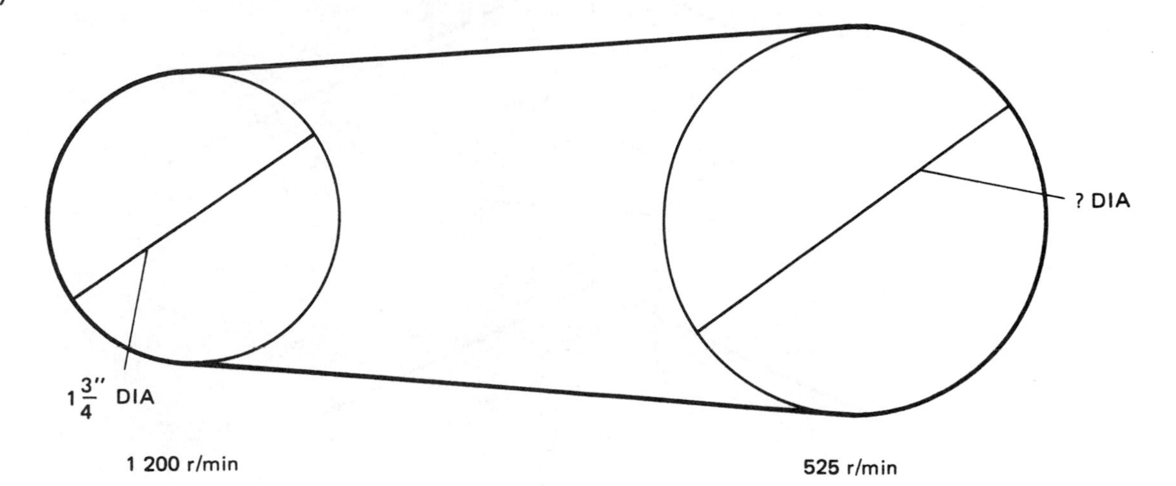

46. This rotary compressor has the piston positioned for maximum volume and for minimum volume. What is the angle between these piston positions?

(24)

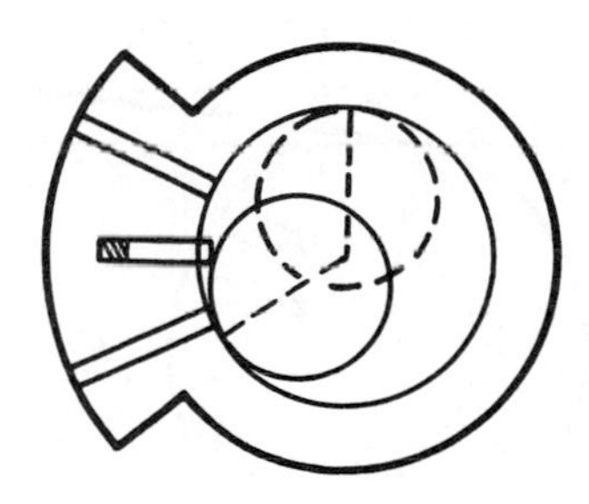

____

47. A triangular-shaped shim is used to level an oil tank. How high will the legs of the tank have to be raised? __________

(36)

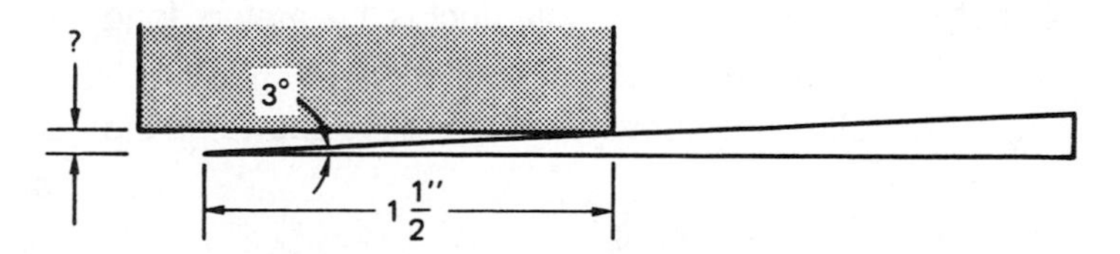

48. To support a duct, a brace is put across the corner of a room. Find, to the nearer whole degree, angle **A**. __________

(36)

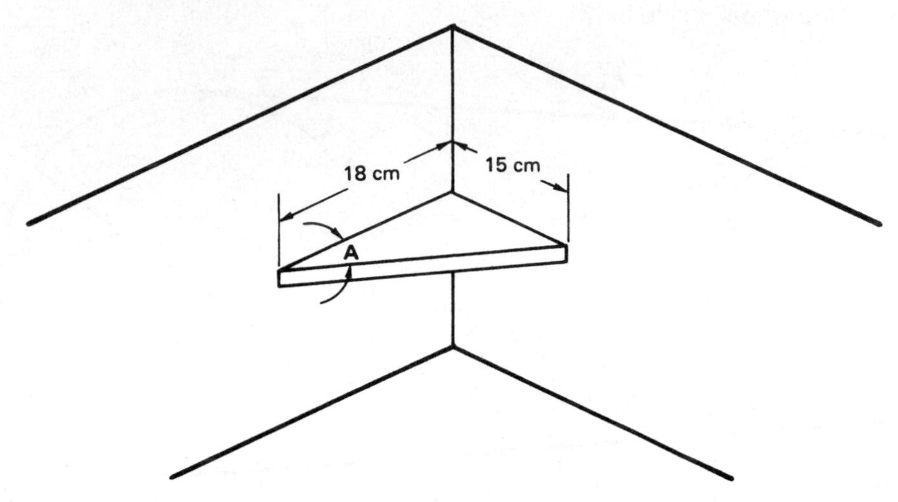

49. Find the length of belt that is in contact with the pulley wheel. Round the answer to the nearer hundredth inch. __________

(35)

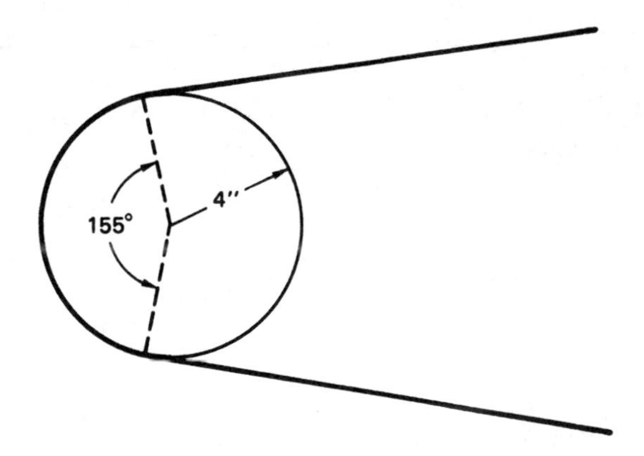

50. A round duct has a diameter of 25 centimeters. The duct is 1.2 meters long and has a grooved seam. The seam allowance is 0.85 centimeter.
    a. Find, in centimeters, the length of the stretchout. Round the answer to the nearer hundredth. a. __________
    b. Find, in meters, the width of the stretchout. b. __________

(34)

ACHIEVEMENT REVIEW C

1. (1) $\begin{array}{r} 5\,086 \\ +\;\;977 \\ \hline \end{array}$

2. (2) $\begin{array}{r} 4\,133 \\ -\,2\,507 \\ \hline \end{array}$

3. (3) $\begin{array}{r} 607 \\ \times\;\;53 \\ \hline \end{array}$

4. (4) $36\overline{)7\,128}$

5. (11) $\begin{array}{r} 44.215 \\ .372\,9 \\ +\,541.2 \\ \hline \end{array}$

6. (12) $\begin{array}{r} 273.42 \\ -\,180.447 \\ \hline \end{array}$

7. (13) $\begin{array}{r} 8.515 \\ \times\,13.07 \\ \hline \end{array}$

8. (14) $.047\overline{)2.975\,1}$

9. (6) $\frac{5}{8} + \frac{2}{3} =$

10. (6) $16\frac{2}{3} + 8\frac{4}{5} =$

11. (7) $\frac{7}{9} - \frac{1}{5} =$

12. (7) $2\frac{5}{16} - 1\frac{1}{2} =$

13. $\frac{3}{4} \times \frac{16}{21} =$
(8)

14. $4\frac{2}{3} \times 5\frac{1}{4} =$
(8)

15. $\frac{7}{8} \div \frac{1}{8} =$
(9)

16. $3\frac{3}{5} \div 4\frac{1}{9} =$
(9)

17. An installation contractor is awarded the bid on a building project of 73 identical apartments. To install the heating system in one apartment, 609 feet of Romex cable was used (including waste). How much wire must the contractor order for the total job? __________
(3)

18. A pulley wheel must be replaced on a motor shaft to run a fan. The shaft diameter is 0.627 inch. Can a pulley with a 5/8-inch hole fit on the shaft? __________
(15)

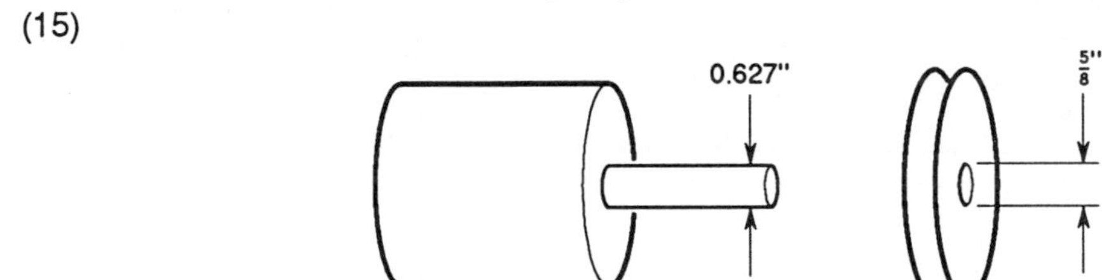

19. A 1 200 gallon fuel oil truck made deliveries of 112.3 gallons, 221.5 gallons, 97.7 gallons, 141.0 gallons, and 193.8 gallons. How much fuel is left on the truck? __________
(12)

20. An installer charging $6.25 per hour sent a bill to a customer for $171.85 for labor after installing a heating system. How long did it take to install the heating system? Round off to the nearer tenth hour. __________
(14)

21. A resistor used in a heating system control circuit has a value of 1 800 ohms ± 5%.
    a. What is the smallest value the resistor can have and be within the tolerance? a. __________
    b. What is the largest value the resistor can have and be within the tolerance? b. __________
(17)

22. During the month of February, one repairer found that 85% of his calls were emergency calls. If he made a total of 80 calls during the month, how many were emergency calls? __________

(17)

23. The Stay Warm Heating Supply Co. charges heating contractors list price minus 8% for supplies. R & B Heating buys duct material listing for $237.00. How much does R & B have to pay for the material? __________

(18)

24. A car air conditioning system takes refrigerant R-12 as low-pressure vapor (31 psig) and compresses it to high-pressure vapor (190 psig). What is the compression ratio of the compressor? (Use absolute pressures.) Round off to the nearer tenth. __________

(19, 31)

25. One hundred pounds of refrigerant R-500 is composed of 73.8 pounds of R-12 and 26.2 pounds of R-152a. What is the ratio of R-12 to R-152a in R-500? Round off to nearer tenth. __________

(19)

26. An 1 800-revolutions-per-minute (rpm) motor is to turn a fan at 2 400 rpm. If the motor's pulley has a diameter of 4 1/2 inches, what diameter pulley should go on the fan? __________

(20)

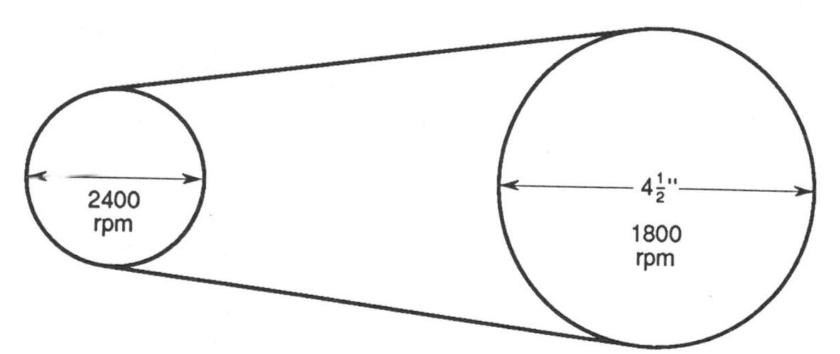

27. Convert the dimensions of this window to inches. ____________
(21)

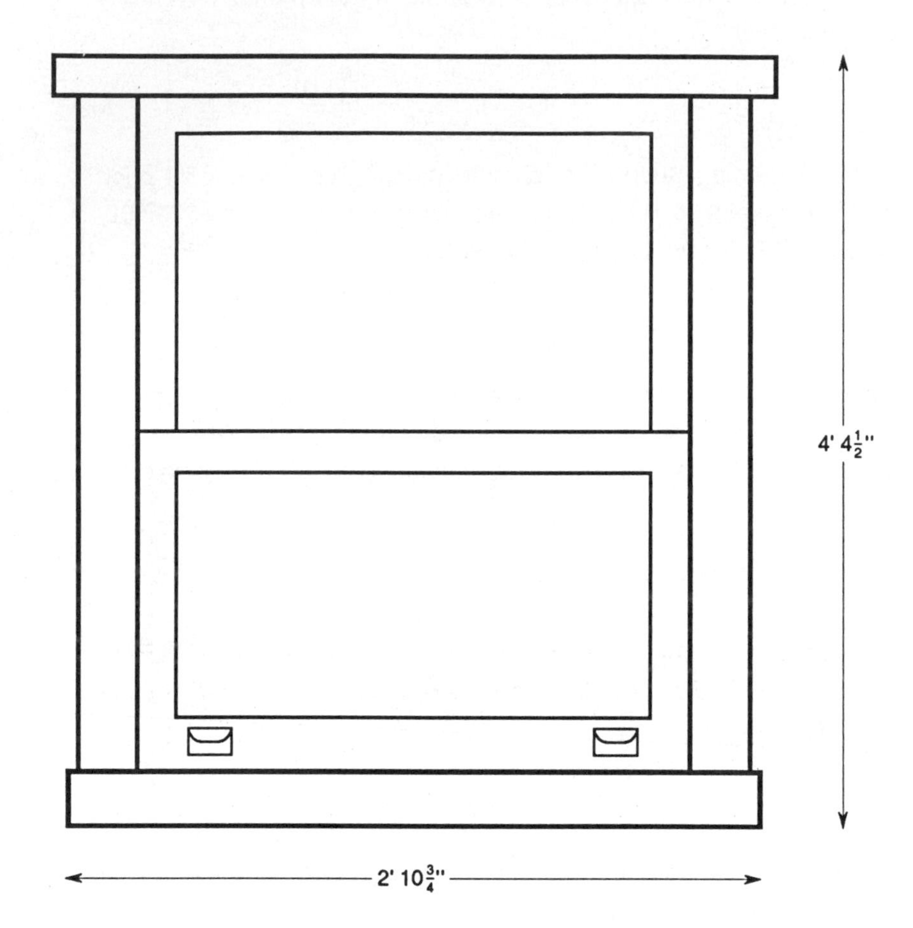

28. Instructions for installing supports for a through-the-wall air conditioner state that 9.5 millimeter holes should be drilled. Would a 3/8-inch drill make a larger or smaller hole than needed? ____________
(22)

29. A heat pump is added to a hot-water heater. The hot-water heater is also wrapped in insulation. What is the length of the insulation around the hot-water heater? __________

(23)

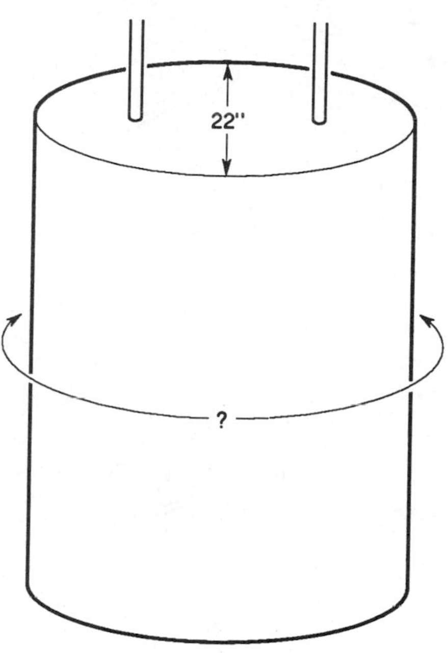

30. Once installed, a through-the-wall room air conditioner is sealed using caulk. How long must the bead of caulk be to seal the air conditioner shown? __________

(23)

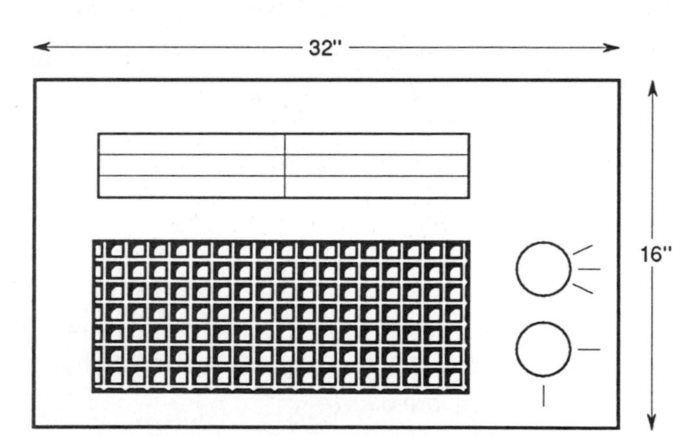

31. A new office building is built as shown. What angle bend must the ducts have to stay along the outside wall? __________

(24)

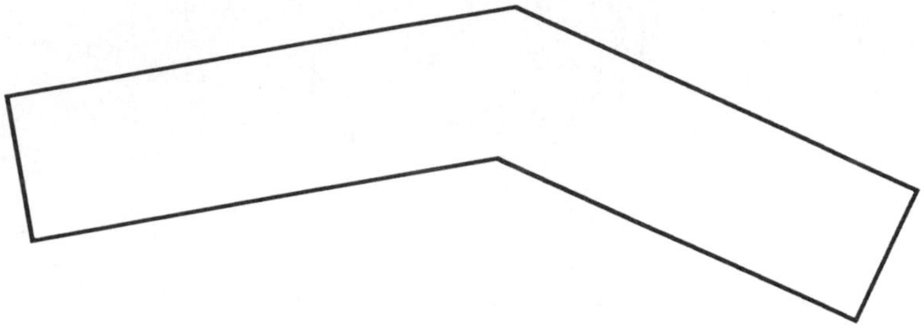

32. Many refrigeration cycles operate between the temperatures 5°F and 86°F. What are these temperatures expressed as degrees Celsius? __________

(25)

33. A rectangular foam filter measures 32.4 centimeters (cm) by 17.9 cm. What is the surface area of this filter in square centimeters? __________

(13, 26)

34. Air flows in a duct. If the cross-sectional area of the duct decreases, the speed of the air will increase. If the cross-sectional area of the duct increases, the speed of the air will decrease.

    Air is flowing through a 6-inch square duct. The duct changes into a 6-inch diameter round duct.

    a. Is the area of the round duct the same as, smaller than, or larger than the square duct? a. __________

    b. Will the air flowing in the round duct be the same speed as, slower than, or faster than the air flowing in the square duct? b. __________

(26)

35. The ceiling in the room shown must be insulated. What is the area in square feet that must be insulated? Round to the nearer tenth square foot. ______________

(27)

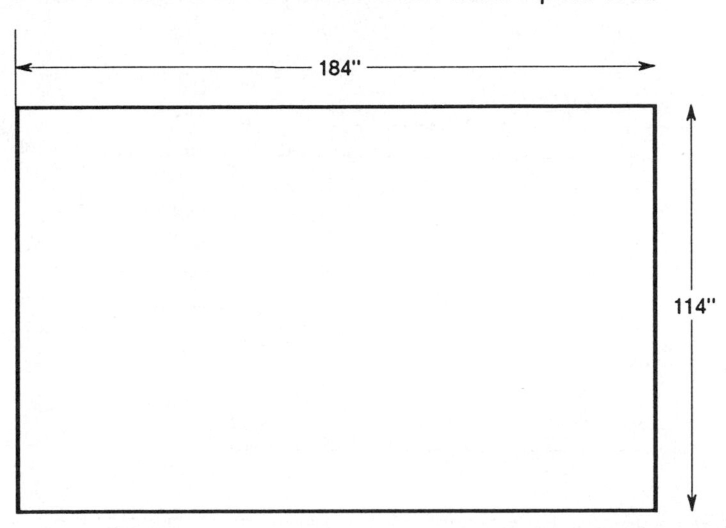

36. The plate shown fits over the end of a heating duct, sealing it. Seven of these plates must be painted.

a. Find the area in square inches that must be painted. Round to the nearer tenth square inch. a. ______________

b. Find this area in square feet. Round to the nearer tenth square foot. b. ______________

(27)

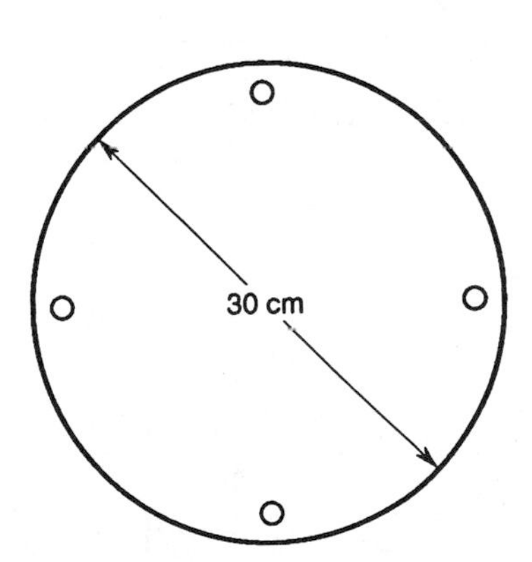

37. A shop area and office need to be air conditioned. The shop has a 14-foot ceiling while the office has an 8-foot ceiling. What volume in cubic feet must be air conditioned? ____________

(28)

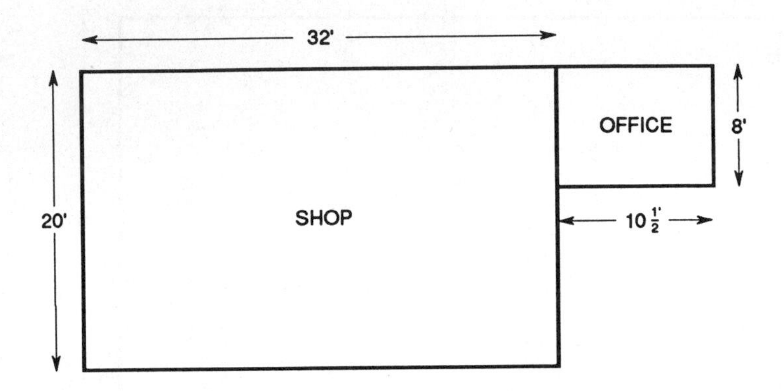

38. A bin to hold rocks that serves as the storage unit for a solar heat collector is 8 feet long, 11 feet wide, and 7 feet high. How many cubic yards of rock are needed to fill the bin? Round to the nearer cubic yard. ____________

(28)

39. A bottled-water water cooler has a tank that needs cooling. The tank is shown. What is the volume of water that gets cooled? (Hint: Subtract the smaller cylindrical volume from the larger cylindrical volume.) ____________

(29)

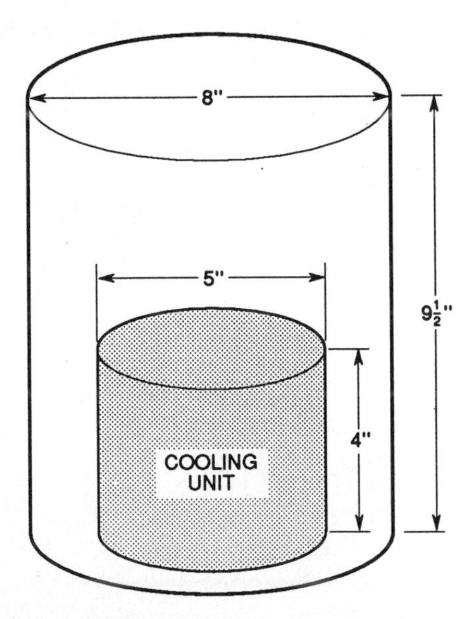

40. A 3 000-watt window air conditioner is wired by itself to a 230-volt circuit. What should be the rating of the circuit breaker to protect the circuit and not trip every time the unit is running? (Circuit breakers are available in 5, 10, 15, 20, and 25 ampere values.) __________

(30)

41. A quartz electric heater is listed for 115 volts. It draws 13 amperes of current. What is the resistance of the heater? Round to the nearer hundredth ohm. __________

(30)

42. A cylinder of R-12 sitting on the back of a truck shows a pressure of 250 pounds per square inch gauge pressure (psig) (264.7 pounds per square inch absolute (psia)) when the outside temperature is 46°F (281 K). Sun shining on the tank heats it to 120°F (322 K). What would the pressure now read? Round to the nearer tenth psig. __________

(31)

43. A refrigerator compressor takes in 1.3 cubic inches of R-12 at 15°F and a pressure of 15 psig on each stroke. It compresses it to a pressure of 175 psig. This raises its temperature to 190°F.
    a. Find the absolute pressures associated with this problem. a. __________
    b. Convert the temperatures in this problem to equivalent ones using the Kelvin scale. Round to the nearer degree. b. __________
    c. What volume does the R-12 vapor now occupy? Round to the nearer tenth cubic inch. c. __________

(31, 25)

44. The end wall of a room measures 12 feet wide and 8 feet high and has no windows. The region where the house was built has a design temperature difference of 70°F. The wall is wood frame with sheathing and siding. How much is the heat load reduced if 1-inch polystyrene sheathing is added to the R-11 insulation already there? (Use heat load values from the table.) __________

(32)

45. A 24-inch-long rectangular duct of dimensions 6 inches by 10 inches is to be built.
    a. What is the width of the stretchout? a. __________
    b. What is the length of the stretchout if it is a welded seam duct? b. __________
    c. What is the length of the stretchout if the duct has a 3/4-inch lap seam? c. __________

(33)

46. A 16-inch-long piece of stovepipe has a 1/4-inch grooved seam. The pipe is a 6-inch diameter pipe.
    a. What is the width of the stretchout? a. ______________
    b. What is the length of the stretchout? Round to the nearer eighth inch. b. ______________

(34)

47. A 16-centimeters-wide rectangular duct must make a 30° bend. The radius of the throat is 8 centimeters.
    a. What is the arc length of the throat to the nearer tenth of a centimeter? a. ______________
    b. What is the arc length of the heel to the nearer tenth of a centimeter? b. ______________

(35)

48. A power cable to a heating unit must be replaced and rerouted. The old cord is 25 feet long. What is the total length of the new cable? ______________

(36)

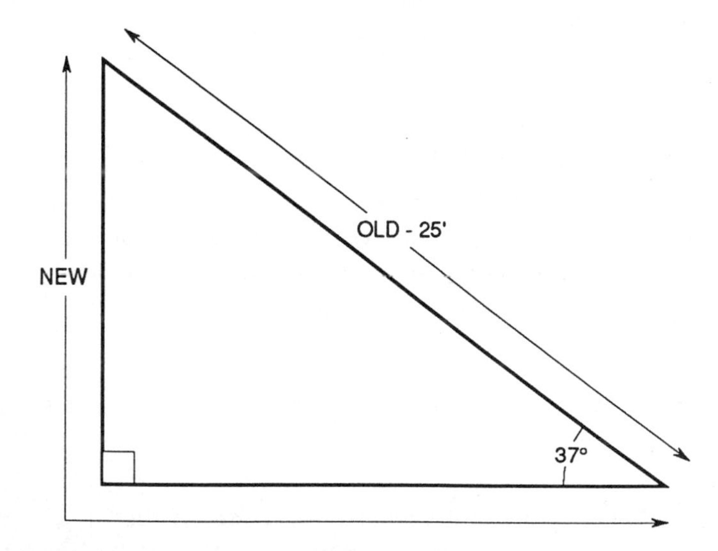

49. How many more repair calls were made in November than May? ____________
(37)

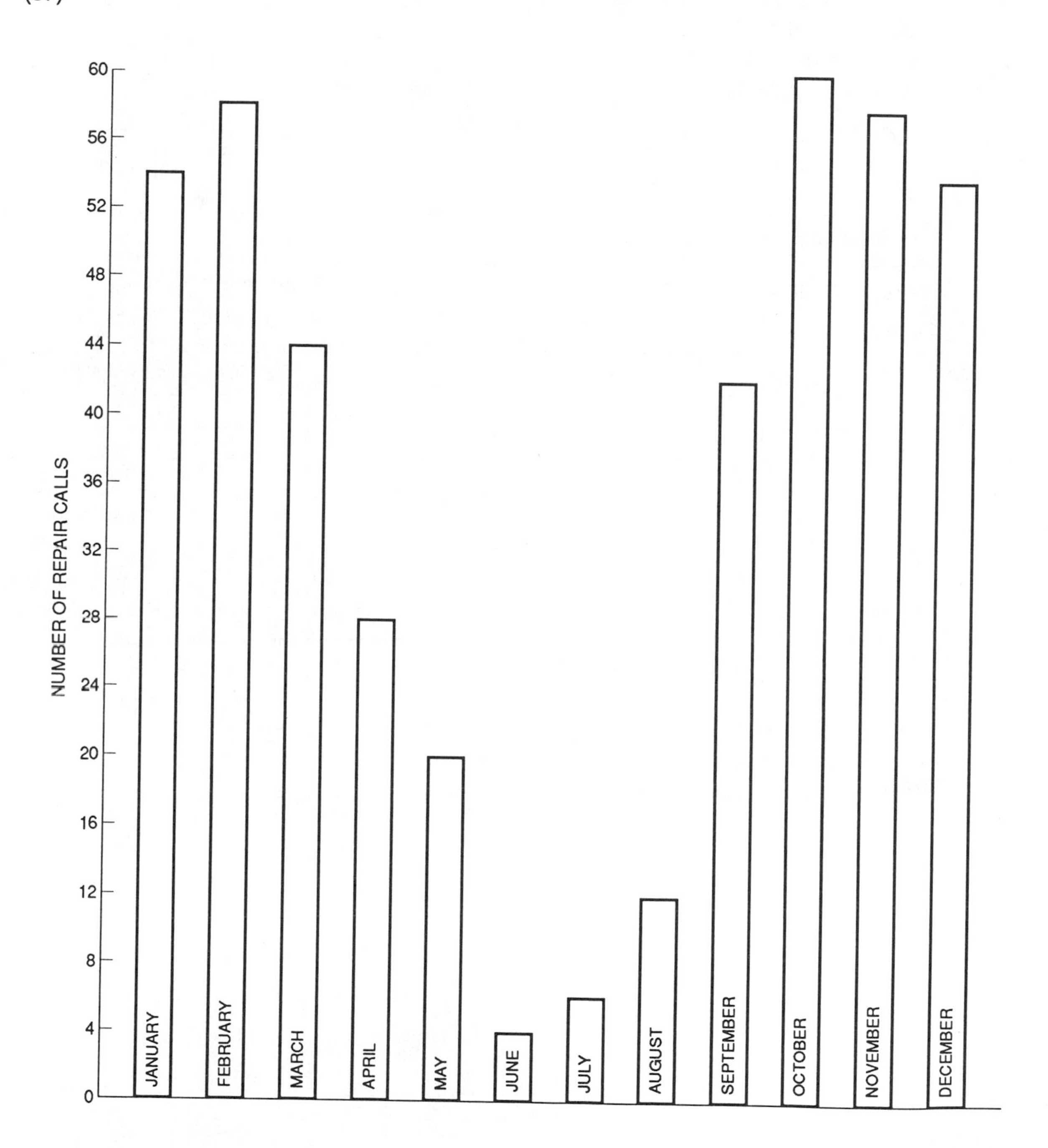

50. Prepare a bill for the following items:
    a. Labor of 17 1/2 hours at $5.75 per hour — a. ______
    b. Repair parts $32.50 — b. ______
    c. Freon recharge $8.35 — c. ______
    d. Service call $17.50 — d. ______
    e. 6% sales tax — e. ______

(38)

Keep Kool Refrigeration Co.
1000 Main St.
Anytown, SC 29000

Customer:

Date
Salesperson
No. 3149

QUANTITY	ITEM	PRICE	AMOUNT
ORIGINAL	**Thank You**		

APPENDIX

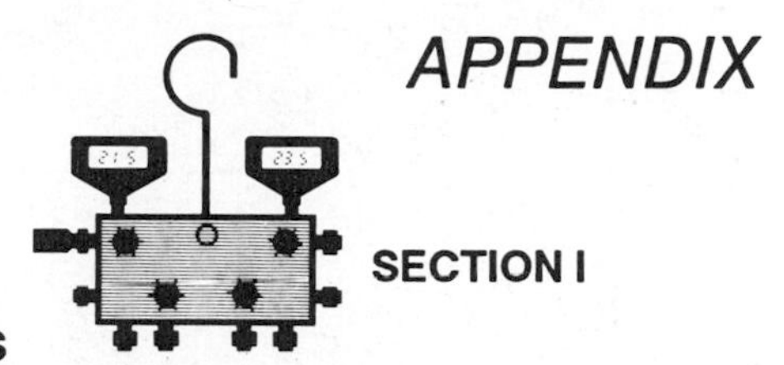

SECTION I

DENOMINATE NUMBERS

Denominate numbers are numbers that include units of measurement. The units of measurement are arranged from the largest units at the left to the smallest unit at the right.

For example: 6 yd 2 ft 4 in

All basic operations of arithmetic can be performed on denominate numbers.

I. EQUIVALENT MEASURES

Measurements that are equal can be expressed in different terms. For example, 12 in = 1 ft If these equivalents are divided, the answer is 1.

$$\frac{1\text{ ft}}{12\text{ in}} = 1 \qquad \frac{12\text{ in}}{1\text{ ft}} = 1$$

To express one measurement as another equal measurement, multiply by the equivalent in the form of 1.

To express 6 inches in equivalent foot measurement, multiply 6 inches by 1 in the form of $\frac{1\text{ft}}{12\text{ in}}$. In the numerator and denominator, divide by a common factor.

$$6\text{ in} = \frac{\overset{1}{\cancel{6\text{ in}}}}{1} \times \frac{1\text{ ft}}{\underset{2}{\cancel{12\text{ in}}}} = \frac{1}{2}\text{ ft or } 0.5\text{ ft}$$

To express 4 feet in equivalent inch measurement, multiply 4 feet by 1 in the form of $\frac{12\text{ in}}{1\text{ ft}}$

$$4\text{ ft} = \overset{4}{\cancel{4\text{ ft}}} \times \frac{12\text{ in}}{\underset{1}{\cancel{1\text{ ft}}}} = \frac{48\text{ in}}{1} = 48\text{ in}$$

Per means division, as with a fraction bar. For example, 50 miles per hour can be written $\frac{50\text{ miles}}{1\text{ hour}}$.

II. BASIC OPERATIONS

A. ADDITION

Sample: 2 yd 1 ft 5 in + 1 ft 8 in + 5 yd 2 ft

1. Write the denominate numbers in a column with like units in the same column.

	2 yd.	1 ft.	5 in.
		1 ft.	8 in.
+	5 yd.	2 ft.	

2. Add the denominate numbers in each column.

7 yd.	4 ft.	13 in.

3. Express the answer using largest possible units.

7 yd.			=	7 yd.		
	4 ft.		=	1 yd.	1 ft.	
		13 in.	= +		1 ft.	1 in.
7 yd.	4 ft.	13 in.	=	8 yd.	2 ft.	1 in.

B. SUBTRACTION

Sample: 4 yd. 3 ft. 5 in. – 2 yd. 1 ft. 7 in.

1. Write the denominate numbers in columns with like units in the same column.

	4 yd.	3 ft.	5 in.
–	2 yd.	1 ft.	7 in.

2. Starting at the right, examine each column to compare the numbers. If the bottom number is larger, exchange one unit from the column at the left for its equivalent. Combine like units.

3. Subtract the denominate numbers.

4. Express the answer using the largest possible units.

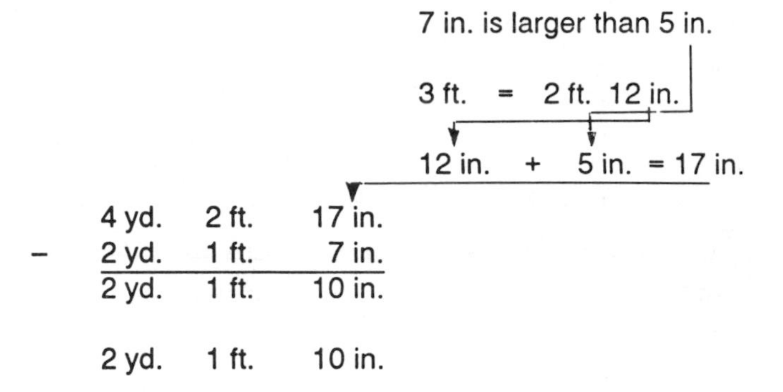

C. MULTIPLICATION
–By a constant

Sample: 1 hr. 24 min. × 3

1. Multiply the denominate number by the constant.

1 hr.	24 min.
	×3
3 hr.	72 min.

2. Express the answer using the largest possible units.

3 hr.		=	3 hr.	
	72 min.	=	1 hr.	12 min.
3 hr.	72 min.	=	4 hr.	12 min.

–By a denominate number expressing linear measurement

Sample: 9 ft 6 in × 10 ft

1. Express all denominate numbers in the same unit. — 9 ft 6 in = $9\frac{1}{2}$ ft

2. Multiply the denominate numbers. (This includes the units of measure, such as ft × ft = sq ft) — $9\frac{1}{2}$ ft × 10 ft = $\frac{19}{2}$ ft × 10 ft = 95 sq ft

–By a denominate number expressing square measurement

Sample: 3 ft × 6 sq ft

1. Multiply the denominate numbers. (This includes the units of measure, such as ft × ft = sq ft and sqft × ft = cu ft) — 3 ft × 6 sq ft = 18 cu ft

–By a denominate number expressing rate

Sample: 50 miles per hour × 3 hours

1. Express the rate as a fraction using the fraction bar for *per*. — $\frac{\text{50 miles}}{\text{1 hour}} \times \frac{\text{3 hours}}{1}$

2. Divide the numerator and denominator by any common factors, including units of measure. — $\frac{\text{50 miles}}{\cancel{\text{1 hour}}_{1}} \times \frac{\cancel{\text{3 hours}}^{3}}{1}$

3. Multiply numerators. Multiply denominators. — $\frac{\text{150 miles}}{1}$ =

4. Express the answer in the remaining unit. — 150 miles

D. DIVISION

–By a constant

Sample: 8 gal 3 qt ÷ 5

1. Express all denominate numbers in the same unit. — 8 gal 3 qt = 35 qt

2. Divide the denominate number by the constant. — 35 qt ÷ 5 = 7 qt
3. Express the answer using the largest possible units. — 7 qt = 1 gal 3 qt

–By a denominate number expressing linear measurement

Sample: 11 ft 4 in ÷ 8 in

1. Express all denominate numbers in the same unit. — 11 ft 4 in = 136 in
2. Divide the denominate numbers by a common factor. (This includes the units of measure, such as inches ÷ inches = 1.) — 136 in ÷ 8 in = $\frac{\overset{17}{\cancel{136\text{ in}}}}{\underset{1}{\cancel{8\text{ in}}}} = \frac{17}{1} = 17$

—By a linear measure with a square measurement as the dividend

Sample: 20 sq ft ÷ 4 ft

1. Divide the denominate numbers. (This includes the units of measure, such as sq ft ÷ ft = ft) — 20 sq ft ÷ 4 ft $\frac{\overset{5\text{ ft}}{\cancel{20\text{ sq ft}}}}{\cancel{4\text{ ft}}} = \frac{5\text{ ft}}{1}$
2. Express the answer in the remaining unit. — 5 ft

—By denominate numbers used to find rate

Sample: 200 mi ÷ 10 gal

1. Divide the denominate numbers — $\frac{\overset{20\text{ mi}}{\cancel{200\text{ mi}}}}{\underset{1\text{ gal}}{\cancel{10\text{ gal}}}} = \frac{20\text{ mi}}{1\text{ gal}}$
2. Express the units with the fraction bar meaning *per*. — $\frac{20\text{ mi}}{1\text{ gal}}$ = 20 miles per gallon

Note: Alternate methods of performing the basic operations will produce the same result. The choice of method is determined by the individual.

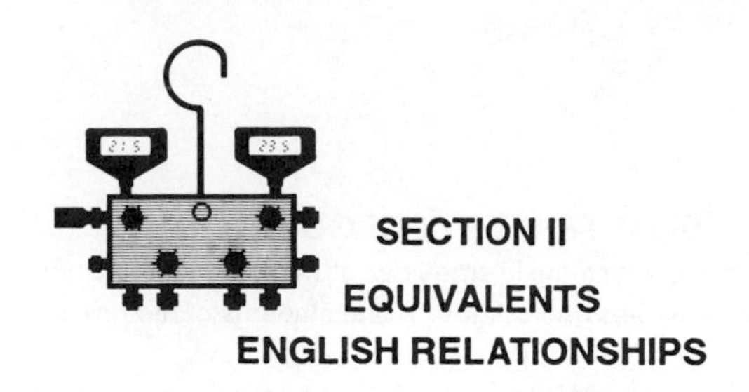

SECTION II
EQUIVALENTS
ENGLISH RELATIONSHIPS

LENGTH EQUIVALENTS

1/16 inch	=	0.062 5 inch
1/8 inch	=	0.125 inch
3/16 inch	=	0.187 5 inch
1/4 inch	=	0.25 inch
5/16 inch	=	0.312 5 inch
3/8 inch	=	0.375 inch
7/16 inch	=	0.437 5 inch
1/2 inch	=	0.5 inch
9/16 inch	=	0.562 5 inch
5/8 inch	=	0.625 inch
11/16 inch	=	0.687 5 inch
3/4 inch	=	0.75 inch
13/16 inch	=	0.812 5 inch
7/8 inch	=	0.875 inch
15/16 inch	=	0.937 5 inch
1 inch	=	$0.083\overline{3}$ foot
2 inches	=	$0.166\overline{6}$ foot
3 inches	=	0.25 foot
4 inches	=	$0.333\overline{3}$ foot
5 inches	=	$0.416\overline{6}$ foot
6 inches	=	0.5 foot
7 inches	=	$0.583\overline{3}$ foot
8 inches	=	$0.666\overline{6}$ foot
9 inches	=	0.75 foot
10 inches	=	$0.833\overline{3}$ foot
11 inches	=	$0.916\overline{6}$ foot

ENGLISH LENGTH MEASURE

1 foot (ft)	=	12 inches (in)
1 yard (yd)	=	3 feet (ft)
1 mile (mi)	=	1 760 yards (yd)
1 mile (mi)	=	5 280 feet (ft)

ENGLISH AREA MEASURE

1 square yard (sq yd)	=	9 square feet (sq ft)
1 square foot (sq ft)	=	144 square inches (sq in)
1 square mile (sq mi)	=	640 acres
1 acre	=	43 560 square feet (sq ft)

ENGLISH VOLUME MEASURE FOR SOLIDS

1 cubic yard (cu yd)	=	27 cubic feet (cu ft)
1 cubic foot (cu ft)	=	1 728 cubic inches (cu in)

ENGLISH VOLUME MEASURE FOR FLUIDS

1 quart (qt)	=	2 pints (pt)
1 gallon (gal)	=	4 quarts (qt)

ENGLISH VOLUME MEASURE EQUIVALENTS

1 gallon (gal)	=	0.133 681 cubic foot (cu ft)
1 gallon (gal)	=	231 cubic inches (cu in)

SI METRICS STYLE GUIDE

SI metrics is derived from the French name Le Système International d'Unités. The metric unit names are already in accepted practice. SI metrics attempts to standardize the names and usages so that students of metrics will have a universal knowledge of the application of terms, symbols, and units.

The English system of mathematics (used in the United States) has always had many units in its weights and measures tables which were not applied to everyday use. For example, the pole, perch, furlong, peck, and scruple are not used often. These measurements, however, are used to form other measurements and it has been necessary to include the measurements in the tables. Including these measurements aids in the understanding of the orderly sequence of measurements greater or smaller than the less frequently used units.

The metric system also has units that are not used in everyday application. Only by learning the lesser-used units is it possible to understand the order of the metric system. SI metrics, however, places an emphasis on the most frequently used units.

In using the metric system and writing its symbols, certain guidelines are followed. For the student's reference, some of the guidelines are listed.

1. In using the symbols for metric units, the first letter is capitalized only if it is derived from the name of a person.

SAMPLE:	UNIT	SYMBOL	UNIT	SYMBOL
	meter	m	Newton	N (named after Sir Isaac Newton)
	gram	g	degree Celsius	°C (named after Anders Celsius)

EXCEPTIONS: The symbol for liter is L. This is used to distinguish it from the number one (1).

2. Prefixes are written with lowercase letters.

SAMPLE:	PREFIX	UNIT	SYMBOL
	centi	meter	cm
	milli	gram	gm
EXCEPTIONS:	PREFIX	UNIT	SYMBOL
	tera	meter	Tm (used to distinguish it from the metric tonne, t)
	giga	meter	Gm (used to distinguish it from gram, g)
	mega	gram	Mg (used to distinguish it from milli, m)

3. Periods are not used in the symbols. Symbols for units are the same in the singular and the plural (no "s" is added to indicate a plural).

SAMPLE: 1 mm *not* 1 mm. 3 mm *not* 3 mms

4. When referring to a unit of measurement, symbols are not used. The symbol is used only when a number is associated with it.

SAMPLE: The length of the room is expressed in meters. *not* The length of the room is expressed in m. (*The length of the room is 25 m* is correct.)

5. When writing measurements that are less than one, a zero is written before the decimal point.

SAMPLE: 0.25 m *not* .25 m

6. Separate the digits in groups of three, counting from the decimal point to the left and to the right. A space is left between the groups of digits.

SAMPLE: 5 179 232 mm *not* 5,179,232 mm 0.566 23 mg *not* 0.56623 mg 1 346.098 7 *not* 1,346,0987 L

A space is also left between the digits and the unit of measure.

SAMPLE: 5 179 232 mm *not* 5 179 232mm

7. Symbols for area measure and volume measure are written with exponents.

SAMPLE: 3 cm^2 *not* 3 sq cm 4 km^3 *not* 4 cu km

8. Metric words with prefixes are accented on the first syllable. In particular, kilometer is pronounced "kill'-o-meter." This avoids confusion with words for measuring devices which are generally accented on the second syllable, such as thermometer (ther-mom'-e-ter).

METRIC RELATIONSHIPS

The base units in SI metrics include the meter and the gram. Other units of measure are related to these units. The relationship between the units is based on powers of ten and uses these prefixes:

kilo (1 000) centi (0.01) milli (0.001)

These tables show the most frequently used units with an asterisk (*).

METRIC LENGTH MEASURE

10 millimeters (mm)*	=	1 centimeter (cm)*
100 centimeters (cm)	=	1 meter (m)*
1 000 meters (m)	=	1 kilometer (km)*

METRIC AREA MEASURE

100 square millimeters (mm^2)	=	1 square centimeter (cm^2)
10 000 square centimeters (cm^2)	=	1 square meter (m^2)
100 000 square meters (m^2)	=	1 square kilometer (km^2)

METRIC VOLUME MEASURE FOR SOLIDS

1 000 cubic millimeters (mm^3)	=	1 cubic centimeter (cm^3)*
1 000 000 cubic centimetrs (cm^3)	=	1 cubic meter (m^3)*
1 000 000 000 cubic meters (cm^3)	=	1 cubic kilometer (km^3)

METRIC VOLUME MEASURE FOR FLUIDS

100 millimeters (mL)*	=	1 centiliter (cL)
100 centiliters (cL)	=	1 liter (L)*
1 000 liters (L)	=	1 kiloliter (kL)

METRIC VOLUME MEASURE EQUIVALENTS

1 000 cubic centimeters (cm^3)	=	1 liter (L)
1 cubic centimeter (cm^3)	=	1 milliliter (L)

METRIC MASS MEASURE

10 milligrams (mg)*	=	1 centigram (cg)
100 centigrams (cg)	=	1 gram (g)*
1 000 grams (g)	=	1 kilogram (kg)*
1 000 kilograms (kg)	=	1 megagram (Mg)*

- ▲ To express a metric length unit as a smaller metric length unit, multiply by a positive power of ten such as 10, 100, 1 000, 10 000, etc.
- ▲ To express a metric length unit as a larger metric length unit, multiply by a negative power of ten such as 0.1, 0.01, 0.001, 0.0001, etc.
- ▲ To express a metric area unit as a smaller metric area unit, multiply by 100, 10 000, 1 000 000, etc.
- ▲ To express a metric area unit as a larger metric area unit, multiply by 0.01, 0.000 1, 0.000 001, etc.
- ▲ To express a metric volume unit for solids as a smaller metric volume unit for solids, multiply by 1 000, 1 000 000, 1 000 000 000, etc.
- ▲ To express a metric volume unit for solids as a larger metric volume unit for solids, multiply by 0.001, 0.000 001, 0.000 000 001, etc.
- ▲ To express a metric volume unit for fluids as a smaller metric volume unit for fluids, multiply by 10, 100, 1 000, 10 000, etc.
- ▲ To express a metric volume unit for fluids as a larger metric volume unit for fluids, multiply by 0.1, 0.01, 0.001, 0.000 1, etc.
- ▲ To express a metric mass unit as a smaller metric mass unit, multiply by 10, 100, 1 000, 10 000, etc.
- ▲ To express a metric mass unit as a larger metric mass unit, multiply by 0.1, 0.01, 0.001, 0.000 1, etc.

Metric measurements are expressed in decimal parts of a whole number. For example, one-half millimeter is written as 0.5 mm.

In calculating with the metric system, all measurements are expressed using the same prefixes. If answers are needed in millimeters, all parts of the problem should be expressed in millimeters before the final solution is attempted. Diagrams that have dimensions in different prefixes must first be expressed using the same unit.

ENGLISH-METRIC EQUIVALENTS

LENGTH MEASURE

1 inch (in)	=	25.4 millimeters (mm)
1 inch (in)	=	2.54 centimeters (cm)
1 foot (ft)	=	0.304 8 meter (m)
1 yard (yd)	=	0.914 4 meter (m)
1 mile (mi)	≈	1.609 kilometers (km)
1 millimeter	≈	0.039 37 inch (in)
1 centimeter (cm)	≈	0.393 7 inch (in)
1 meter (m)	≈	3.280 84 feet (ft)
1 meter (m)	≈	1.093 61 yards (yd)
1 kilometer (km)	≈	0.621 37 mile (mi)

AREA MEASURE

1 square inch (sq in)	=	645.16 square millimeters (mm^2)
1 square inch (sq in)	=	6.451 6 square centimeters (cm^2)
1 square foot (sq ft)	≈	0.092 903 square meter (m^2)
1 square yard (sq yd)	≈	0.836 127 square meter (m^2)
1 square millimeter (mm^2)	≈	0.001 55 square inch (sq in)
1 square centimeter (cm^2)	≈	0.155 square inch (sq in)
1 square meter (m^2)	≈	10.763 91 square feet (sq ft)
1 square meter (m^2)	≈	1.195 99 square yards (sq yd)

VOLUME MEASURE FOR SOLIDS

1 cubic inch (cu in)	=	16.387 064 cubic centimeters (cm^3)
1 cubic foot (cu ft)	≈	0.028 317 cubic meter (m^3)
1 cubic yard (cu yd)	≈	0.764 555 cubic meter (m^3)
1 cubic centimeter (cm^3)	≈	0.061 024 cubic inch (cu in)
1 cubic meter (m^3)	≈	35.314 667 cubic feet (cu ft)
1 cubic meter (m^3)	≈	1.307 951 cubic yards (cu yd)

VOLUME MEASURE FOR FLUIDS

1 gallon (gal)	≈	3 785.411 cubic centimeters (cm^3)
1 gallon (gal)	≈	3.785 411 liters (L)
1 quart (qt)	≈	0.946 353 liter (L)
1 ounce (oz)	≈	29.573 53 cubic centimeters (cm^3)
1 cubic centimeter (cm^3)	≈	0.000 264 gallon (gal)
1 liter (L)	≈	0.264 172 gallon (gal)
1 liter (L)	≈	1.056 688 quarts (qt)
1 cubic centimeter (cm^3)	≈	0.033 814 ounce (oz)

MASS MEASURE

1 pound (lb)	≈	0.453 592 kilogram (kg)
1 pound (lb)	≈	453.592 37 grams (g)
1 ounce (oz)	≈	28.349 523 grams (g)
1 ounce (oz)	≈	0.028 35 kilogram (kg)
1 kilogram (kg)	≈	2.204 623 pounds (lb)
1 gram (g)	≈	0.002 205 pound (lb)
1 kilogram (kg)	≈	35.273 962 ounces (oz)
1 gram (g)	≈	0.035 274 ounce (oz)

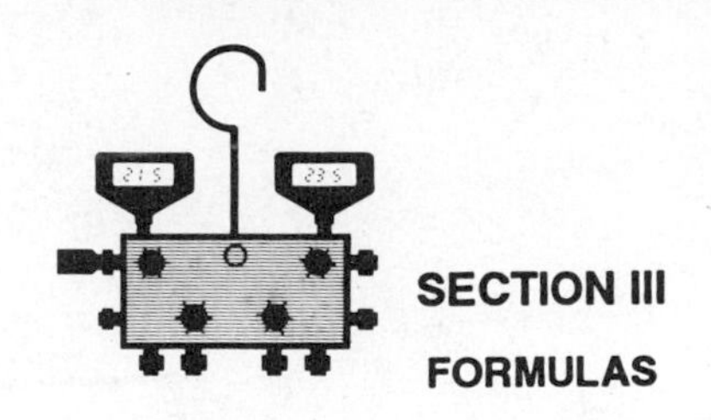

SECTION III
FORMULAS

PERIMETER

Square

$P = 4s$

P = perimeter
s = side

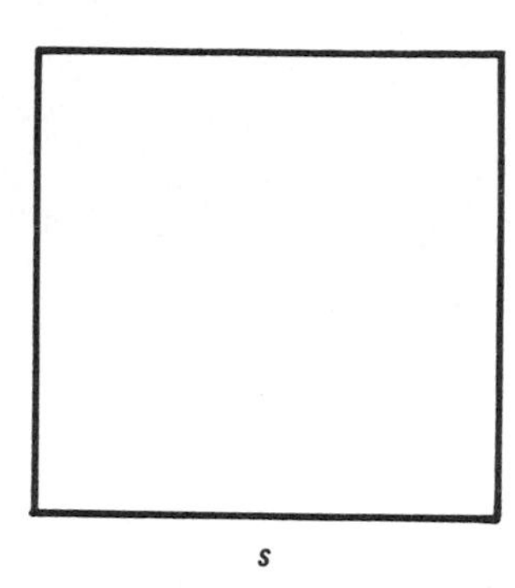

Rectangle

$P = 2l + 2w$

P = perimeter
l = length
w = width

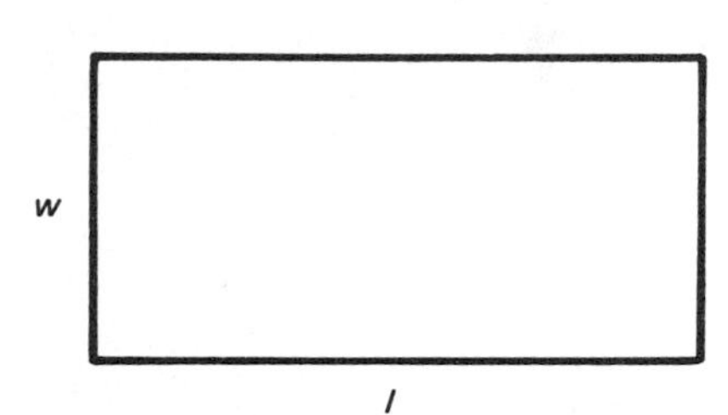

Triangle

$P = a + b + c$

P = perimeter
a = first side
b = second side
c = third side

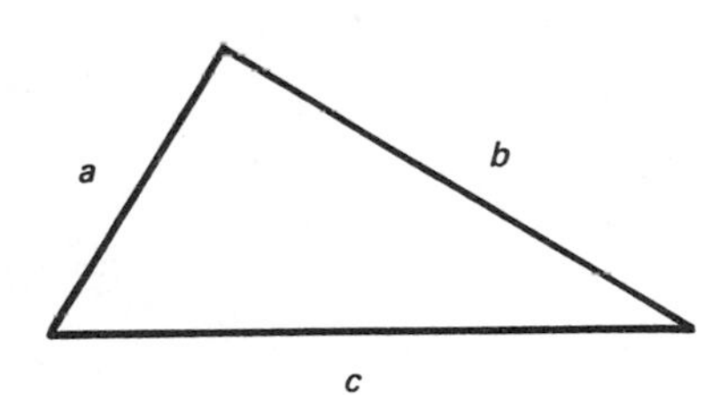

Circle

$C = \pi D$
$C = 2\pi r$

C = *circumference*
π = 3.141 6
D = diameter
r = radius

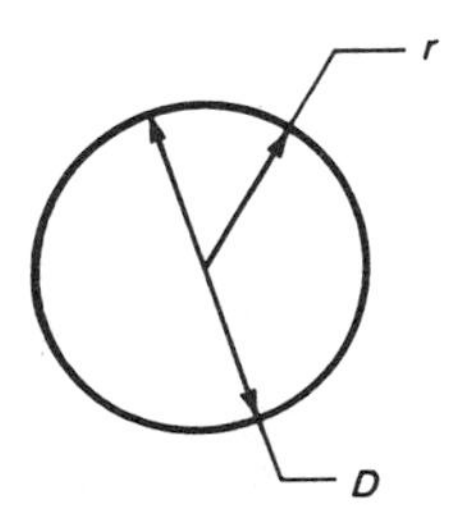

AREA

Square

$A = s^2$

A = area
s = side

Rectangle

$A = lw$

A = area
l = length
w = width

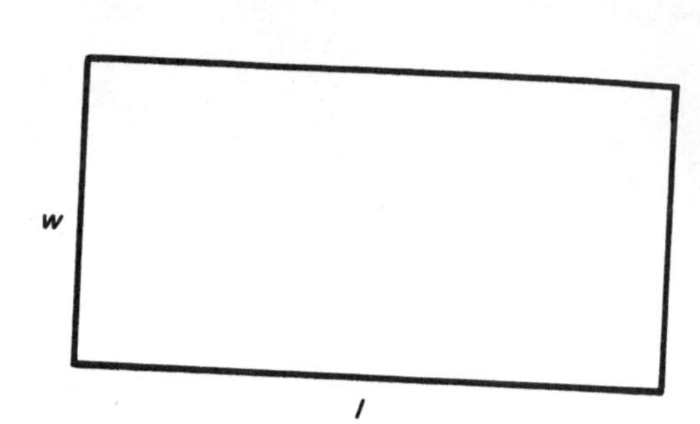

Triangle

$A = \frac{1}{2}bh$

A = area
b = base
h = height

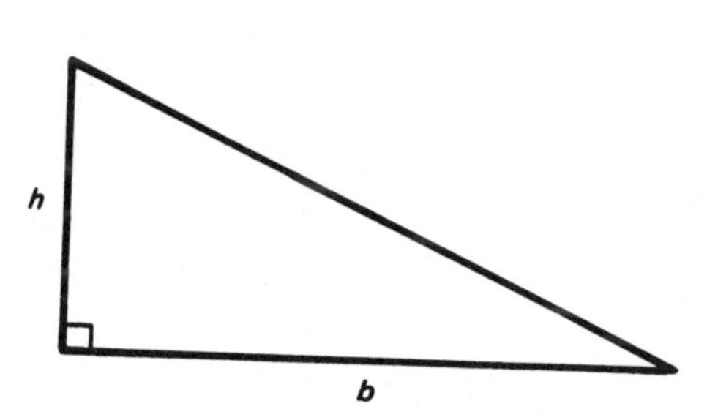

Trapezoid

$A = \frac{1}{2}(b_1 + b_2)h$

A = area
b_1 = first base
b_2 = second base
h = height

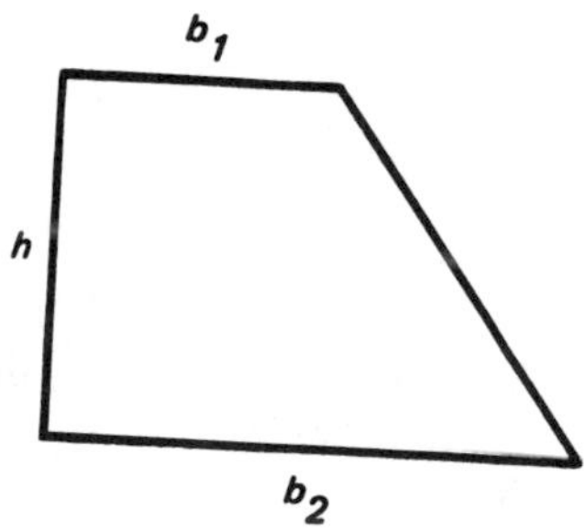

Circle

$A = \pi r^2$

$A = \frac{\pi}{4} D^2$

A = area
π = 3.141 6
r = radius
D = diameter

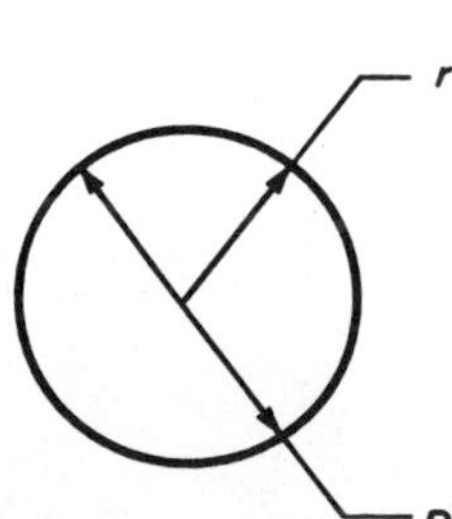

VOLUME

Rectangular Solid

$V = lwh$

V = volume
l = length
w = width
h = height

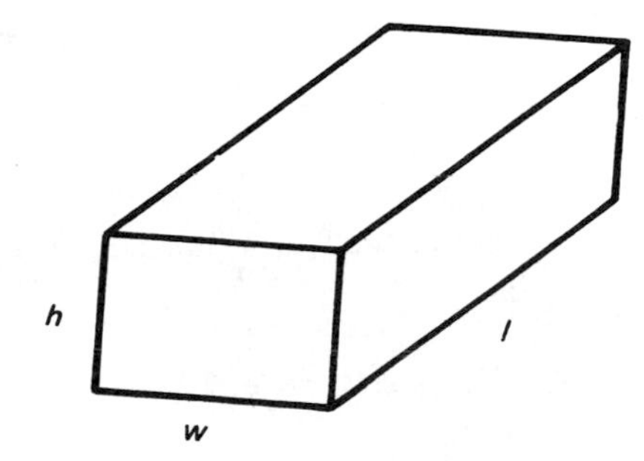

Cylindrical Solid

$V = \pi r^2 h$

$V = \frac{\pi}{4} D^2 h$

V = volume
π = 3.141 6
r = radius
D = diameter
h = height

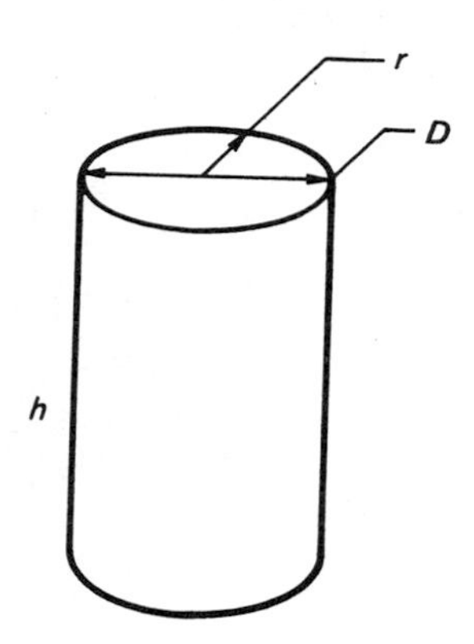

TEMPERATURE

$°C = \frac{5}{9}(°F - 32)$ — °C = *degrees Celsius*

$°F = \frac{9}{5}(°C) + 32$ — °F = *degrees Fahrenheit*

$K = °C + 273$ — K = Kelvins

ELECTRICAL

Ohm's Law

$E = IR$

E = voltage (in volts)
I = current (in amperes)
R = resistance (in ohms)

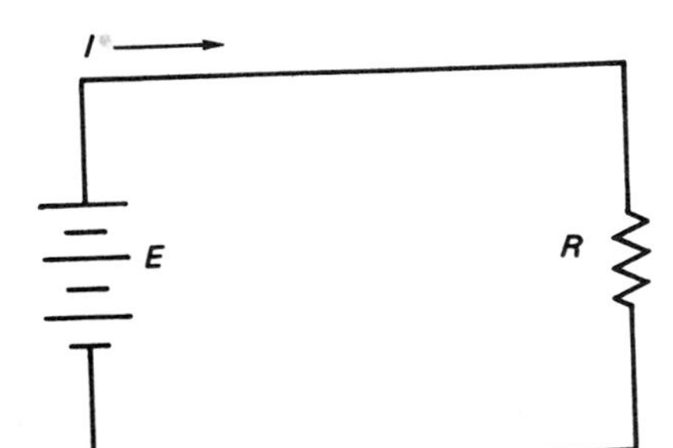

Power Formula

$P = IE$

P = power (in watts)
I = current (in amperes)
E = voltage (in volts)

Resistance in Series

$R = R_1 + R_2 + \ldots$ R = resistance (in ohms)

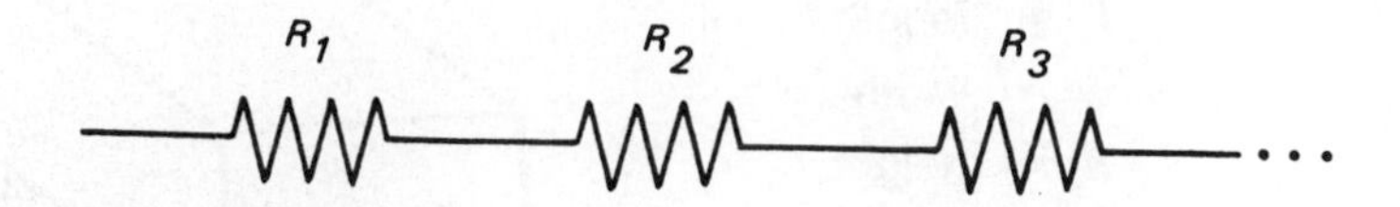

Resistance in Parallel

$$R = \frac{1}{\frac{1}{R_1} + \frac{1}{R_2} + \ldots}$$ R = resistance (in ohms)

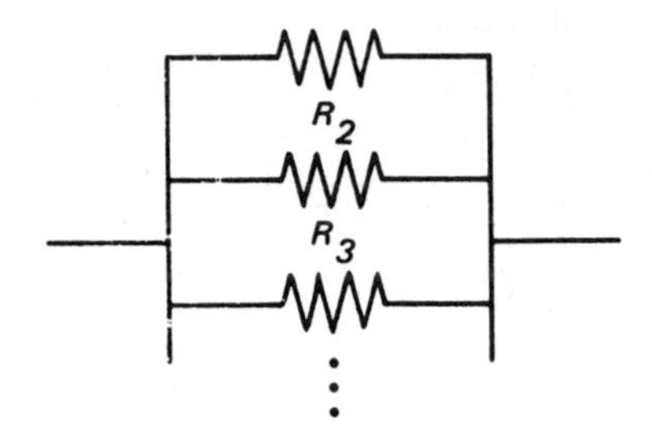

Capacitance in Series

$$C = \frac{1}{\frac{1}{C_1} + \frac{1}{C_2} + \ldots}$$ C = capacitance (in microfarads)

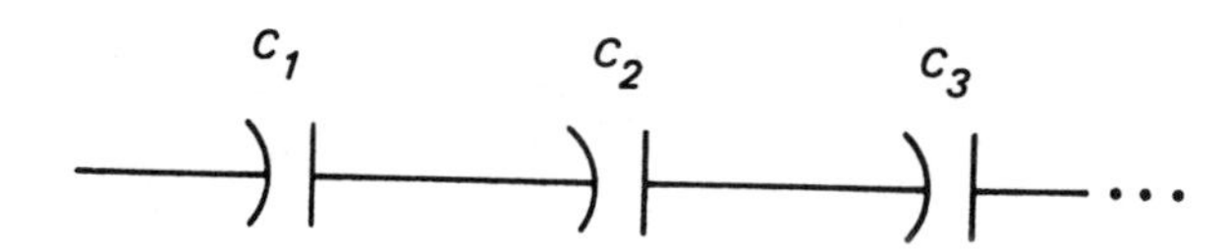

Capacitance in Parallel

$C = C_1 + C_2 + \ldots$ C = capacitance (in microfarads)

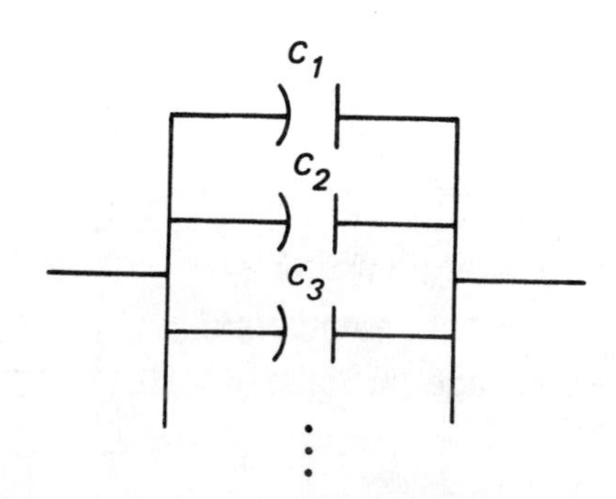

Gas Law

Boyle's Law

$P_1V_1 = P_2V_2$

P = pressure (absolute)
V = volume

Charles' Law

$$\frac{P_1}{T_1} = \frac{P_2}{T_2}$$

P = pressure (absolute)
T = temperature (absolute)

General Gas Law

$$\frac{P_1V_1}{T_1} = \frac{P_2V_2}{T_2}$$

P = pressure (absolute)
V = volume
T = temperature (absolute)

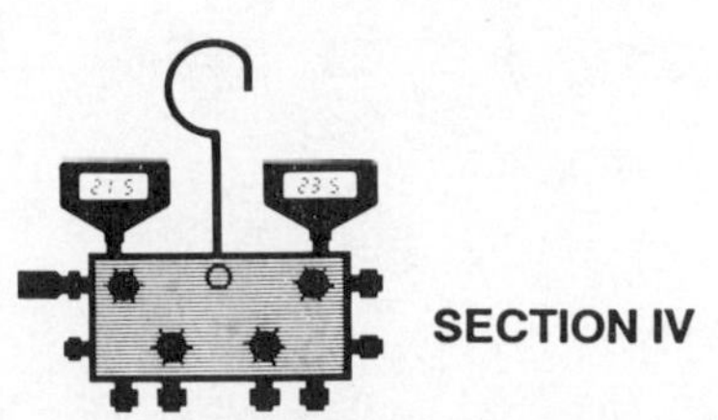

SECTION IV

TRIGONOMETRIC FUNCTIONS

Angle	Sine	Cosine	Tangent	Angle	Sine	Cosine	Tangent
1°	0.017 5	0.999 8	0.017 5	46°	0.719 3	0.694 7	1.035 5
2°	0.034 9	0.999 4	0.034 9	47°	0.731 4	0.682 0	1.072 4
3°	0.052 3	0.998 6	0.052 4	48°	0.743 1	0.669 1	1.110 6
4°	0.069 8	0.997 6	0.069 9	49°	0.754 7	0.656 1	1.150 4
5°	0.087 2	0.996 2	0.087 5	50°	0.766 0	0.642 8	1.191 8
6°	0.104 5	0.994 5	0.105 1	51°	0.777 1	0.629 3	1.234 9
7°	0.121 9	0.992 5	0.122 8	52°	0.788 0	0.615 7	1.279 9
8°	0.139 2	0.990 3	0.140 5	53°	0.798 6	0.601 8	1.327 0
9°	0.156 4	0.987 7	0.158 4	54°	0.809 0	0.587 8	1.376 4
10°	0.173 6	0.984 8	0.176 3	55°	0.819 2	0.573 6	1.428 1
11°	0.190 8	0.981 6	0.194 4	56°	0.829 0	0.559 2	1.482 6
12°	0.207 9	0.978 1	0.212 6	57°	0.838 7	0.544 6	1.539 9
13°	0.225 0	0.974 4	0.230 9	58°	0.848 0	0.529 9	1.600 3
14°	0.241 9	0.970 3	0.249 3	59°	0.857 2	0.515 0	1.664 3
15°	0.258 8	0.965 9	0.267 9	60°	0.866 0	0.500 0	1.732 1
16°	0.275 6	0.961 3	0.286 7	61°	0.874 6	0.484 8	1.804 0
17°	0.292 4	0.956 3	0.305 7	62°	0.882 9	0.469 5	1.880 7
18°	0.309 0	0.951 1	0.324 9	63°	0.891 0	0.454 0	1.962 6
19°	0.325 6	0.945 5	0.344 3	64°	0.898 8	0.438 4	2.050 3
20°	0.342 0	0.939 7	0.364 0	65°	0.906 3	0.422 6	2.144 5
21°	0.358 4	0.933 6	0.383 9	66°	0.913 5	0.406 7	2.246 0
22°	0.374 6	0.927 2	0.404 0	67°	0.920 5	0.390 7	2.355 9
23°	0.390 7	0.920 5	0.424 5	68°	0.927 2	0.374 6	2.475 1
24°	0.406 7	0.913 5	0.445 2	69°	0.933 6	0.358 4	2.605 1
25°	0.422 6	0.906 3	0.466 3	70°	0.939 7	0.342 0	2.747 5
26°	0.438 4	0.898 8	0.487 7	71°	0.945 5	0.325 6	2.904 2
27°	0.454 0	0.891 0	0.509 5	72°	0.951 1	0.309 0	3.077 7
28°	0.469 5	0.882 9	0.531 7	73°	0.956 3	0.292 4	3.270 9
29°	0.484 8	0.874 6	0.554 3	74°	0.961 3	0.275 6	3.487 4
30°	0.500 0	0.866 0	0.577 4	75°	0.965 9	0.258 8	3.732 1
31°	0.515 0	0.857 2	0.600 9	76°	0.970 3	0.241 9	4.010 8
32°	0.529 9	0.848 0	0.624 9	77°	0.974 4	0.225 0	4.331 5
33°	0.544 6	0.838 7	0.649 4	78°	0.978 1	0.207 9	4.704 6
34°	0.559 2	0.829 0	0.674 5	79°	0.981 6	0.190 8	5.144 6
35°	0.573 6	0.819 2	0.700 2	80°	0.984 8	0.173 6	5.671 3
36°	0.587 8	0.809 0	0.726 5	81°	0.987 7	0.156 4	6.313 8
37°	0.601 8	0.798 6	0.753 6	82°	0.990 3	0.139 2	7.115 4
38°	0.615 7	0.788 0	0.781 3	83°	0.992 5	0.121 9	8.144 3
39°	0.629 3	0.777 1	0.809 8	84°	0.994 5	0.104 5	9.514 4
40°	0.642 8	0.766 0	0.839 1	85°	0.996 2	0.087 2	11.430 1
41°	0.656 1	0.754 7	0.869 3	86°	0.997 6	0.069 8	14.300 7
42°	0.669 1	0.743 1	0.900 4	87°	0.998 6	0.052 3	19.081 1
43°	0.682 0	0.731 4	0.932 5	88°	0.999 4	0.034 9	28.636 3
44°	0.694 7	0.719 3	0.965 7	89°	0.999 8	0.017 5	57.290 0
45°	0.707 1	0.707 1	1.000 0	90°	1.000 0	0.000 0	

HEAT TRANSFER MULTIPLIERS

Note: The Heat Transfer Multiplier is found by multiplying the U factor (the amount of heat transferred through 1 square foot of structure for each degree temperature difference between the inside and outside surfaces) by the design temperature difference. The units for the Heat Transfer Multiplier are British thermal units per hour per square foot.

TYPE OF STRUCTURE	DESIGN TEMPERATURE DIFFERENCE		
	25°F	70°F	75°F
Walls — wood frame with sheathing and siding or other veneer			
3 1/2 inches insulation (R-11)	3.5	5	5
3 1/2 inches insulation + 1 inch polystyrene sheathing	3.1	3.5	3.8
Ceiling — under vented roof			
3 1/2 inches insulation (R-11)	2.5	6	6
6 inches insulation (R-19)	1.5	4	4
9 1/2 inches insulation (R-30)	1.0	2.2	2.4
Floor			
No insulation	5	16	17
6 inches insulation	1	3.2	3.4
Windows			
Single pane	35	105	110
Double pane	25	70	75
Single pane + storm window	25	60	65
Double pane (fixed)	25	60	65
Doors			
Insulated core, weather-stripped	5.3	81	86
Sliding glass door, double glass	25	90	95

GLOSSARY

Absolute pressure—The total pressure consisting of the gauge pressure plus the atmospheric pressure.

Absolute temperature—The temperature measured on an absolute scale which uses zero as the temperature at which all molecular motion stops.

Ampere—The unit of electric current.

Atmospheric pressure—The pressure exerted by the gases in the atmosphere upon the earth.

Bending radius—The radius of the circle formed by the arc made when tubing is bent.

British thermal unit—A unit for a quantity of heat. It is the quantity of heat needed to raise the temperature of one pound of water one degree Fahrenheit.

Capacitor—An electrical device used to store electric charges.

Combustion chamber—The part of a furnace where the fuel is burned.

Compression ratio—The ratio of the pressure of the gas leaving a compressor to the pressure of the gas going into the compressor.

Compressor—A device used in refrigerators which takes the refrigerant gas at low pressure and compresses or squeezes it into a gas at high pressure.

Condenser—The part of the refrigerator which takes the hot compressed gas from the compressor and allows it to cool to a liquid.

Condensing pressure—The pressure for a given temperature at which the gas becomes a liquid.

Conduit—A tube used to carry and protect electrical wires.

Connecting rod—The part of a compressor which connects the piston with the crankshaft.

Damper—A valve for controlling airflow.

Density—A measure of the compactness of a substance. It is the mass of the substance per unit volume.

Diffuser—A device for deflecting or spreading out the airflow.

Duct—A tube or channel through which air is moved.

Elbow—A section of a pipe or duct which is bent at an angle.

Electrodes—Parts of an oil burner gun. Two pieces of metal set so there is a gap between them. An electric spark jumps across the gap and ignites the oil.

Evaporator—The part of a refrigerator where the liquid refrigerant vaporizes. Heat is absorbed as the liquid changes to a gas.

Felt wiper—A piece of felt which is in contact with a rotating shaft. The felt is impregnated with oil to lubricate the turning shaft.

Filter-drier—A device for removing small particles and water from the refrigerant.

Fin comb—A device which looks like a comb and is used to straighten the bent metal fins on condensers or evaporators.

Flaring cone—A device used for expanding the end of tubing when making connections.

Flue—A passageway used to carry away smoke and burned gases from a furnace.

Gasket—A device or material used to form a seal between two objects when they are joined.

Gauge pressure—The pressure measured by and read from a gauge. It is the difference between the inside pressure and atmospheric pressure.

Heat load—The amount of heat which must be removed from or added to a space in a given amount of time.

Heat loss—The amount of heat that flows through a boundary from a heated space to an unheated space.

Heat pump—A reversible system which can heat or cool a space.

High pressure side—The portion of a refrigeration system where the refrigerant is under high pressure. This portion extends from the exhaust of the compressor through the condenser to the expansion valve.

Humidifier—A device which puts moisture into the air.

Latent heat of fusion—The amount of heat energy given off as a liquid becomes a solid with no change in temperature or pressure. This is also the amount of heat energy absorbed as the solid becomes a liquid.

Latent heat of vaporization—The amount of heat energy absorbed as a liquid becomes a gas with no change in temperature or pressure. This is also the amount of heat energy given off as the gas becomes a liquid.

Low pressure side—The portion of a refrigeration system where the refrigerant is at low pressure. This portion extends from the expansion valve through the evaporator to the intake of the compressor.

Microfarad—The unit used to measure the electric capacitance.

Motor bearing—A support for the rotating part of a motor.

Mullion heater—The electrical heating element mounted in the part of the refrigerator between the two doors (mullion). The heater is used to prevent sweating.

Parallel circuit—An electric circuit connected in such a manner that the current flows through at least two paths.

Piston—The part of a compressor or motor which moves up and down inside the cylinder.

Piston ring—A thin, narrow ring which fits in a groove around the piston and makes a very tight fit inside the cylinder.

Plenum—A chamber for moving air. A large duct.

Pressure gauge—A device for measuring the pressure of a gas by comparing it with atmospheric pressure.

PVC tubing—Plastic tubing made of polyvinylchloride.

R value—A measure of the resistance to heat flow through a material.

Reed value—A valve used in a compressor. The valve is a flat piece of metal which is placed over the opening. One end of the metal strip is fastened and the other end is free to move up or down to open or close the opening.
Refrigerant—The substance used in a refrigerating system which changes from a gas to a liquid and back again and in doing so transfers heat energy.
Relay capacitor—A relay is a switching device which uses a small electrical signal to activate the switch. The capacitor is the storage device which causes the relay to activate when it becomes totally charged.
Resistance heater—A type of heater which produces heat by allowing a current to flow through a special wire.
Rotary compressor—A device which compresses gases using a rotating motion rather than a reciprocating (up and down) motion.
Rotor—The part that spins in a motor or rotary compressor.
Running capacitor—The storage device which provides a properly timed current to a motor while it is running. The current must be timed properly to cause the motor to turn.
Saturated vapor pressure—The pressure for a given temperature at which a refrigerant can exist as both a vapor and a liquid.
Series circuit—An electric circuit connected in such a manner that the current must flow in a single path.
Shim—A piece of uniform size material such as a washer or a varying thickness material such as a wedge used for filling space.
Smoke pipe—A thin-walled sheet metal tubing used in chimneys.
Spreader—A diffuser which increases the width of an airflow.
Stator—The stationary part of an electric motor.
Strap—A strip of metal used to support a suspended pipe or duct.
Stretchout—The patten for a duct or piece of metal. The pattern is used to cut a flat piece of sheet metal which is then bent to form a duct.
Tee—A joint in pipes or ducts in the shape of the letter T.
Ton refrigeration unit—A unit measuring the amount of refrigeration. One ton of refrigeration will remove in 24 hours the amount of heat needed to melt one ton of ice.
U factor—A measure of the amount of heat flow through a barrier.
V belt—A belt which often connects motors and fans. The belt is named for the V-shaped contact surface.
Vaporizing pressure—A pressure for a given temperature at which a liquid can become a gas.
Volt—The unit of electric potential or electromotive force.
Watt—The unit of electric power.
Weather stripping—A strip of material used to seal joints which do not fit tightly together such as doors or windows.
Y—A joint in pipes or ducts in the shape of the letter Y.

ODD NUMBER ANSWERS

SECTION 1 WHOLE NUMBERS

UNIT 1 ADDITION OF WHOLE NUMBERS

1. 976
3. 8 979
5. 990 m
7. 879
9. 7 298
11. 11 745 ft
13. 46 ft
15. 29 ft
17. 25 ft
19. 9 056 cu ft
21. 46 ft
23. 555 lb
25. 36 442 mi

UNIT 2 SUBTRACTION OF WHOLE NUMBERS

1. 64
3. 711
5. 1 269 qt
7. 331
9. 715
11. 589 cm^2
13. 39°F
15. 18 lb
17. 124 cu ft/min
19. 6 A
21. 15 in
23. 8 straps
25. 40 cu ft/min

UNIT 3 MULTIPLICATION OF WHOLE NUMBERS

1. 219
3. 467 372
5. 17 232 mm
7. 1 235
9. 46 828
11. 483 426 sq ft
13. 1 008 connectors
15. $153
17. 4 932 lb
19. 962 000 Btu
21. 1 825 ft
23. 210 ft
25. 1 932 man hours

UNIT 4 DIVISION OF WHOLE NUMBERS

1. 23
3. 26
5. 356 Btu
7. 43
9. 67
11. 36 lb
13. 83 lb/compressor
15. 475 lb
17. 45 rolls
19. 13 days
21. 14 sheets
23. 725 cu ft
25. 23 cars

UNIT 5 COMBINED OPERATIONS WITH WHOLE NUMBERS

1. 1 873
3. 425
5. 3 348 660
7. 24
9. 5 937 in
11. 27 848 634 s
13. 1 622 cu ft
15. 76 valves
17. a. 12 min
    b. 5 times
19. 215 cu ft/min
21. 38 000 Btu/h
23. 25 ducts
25. a. 288 000 Btu
    b. 12 000 Btu

SECTION 2 COMMON FRACTIONS

UNIT 6 ADDITION OF COMMON FRACTIONS

1. 2 1/5
3. 1 12/25
5. 45 11/16
7. 1 31/32 in
9. 18 5/8 in
11. 3 3/8 in.
13. 22 1/16 in
15. 140 3/4 in
17. 31 2/3 ft
19. 17 1/4 h
21. 27 13/16 in

UNIT 7 SUBTRACTION OF COMMON FRACTIONS

1. 1/8
3. 2 7/15
5. 3 21/80
7. 5/32 in
9. 5 3/8 in
11. 1 3/4 in
13. 5 3/4 h
15. a. 7 3/8 in
    b. 8 5/8 in
17. a. 5 3/4 in
    b. 9 3/4 in
19. 6 8/15 lb

UNIT 8 MULTIPLICATION OF COMMON FRACTIONS

1. 3/16
3. 8/15
5. 2 4/5
7. 1 4/5
9. 7/18 ft
11. 2 3/8 in
13. $990
15. 21 1/2 in
17. 256 1/2 in
19. 30¢
21. a. 4 1/2 h
    b. 3 h
    c. 1/2 h
    d. 1 h

UNIT 9 DIVISION OF COMMON FRACTIONS

1. 18/35
3. 1 4/5
5. 49/78
7. 1/16 in
9. 1 5/7 w
11. 14 straps
13. 12 fins/in
15. 7 pieces
17. a. 23/36 in
    b. 13/20 in
19. 9 seams

UNIT 10 COMBINED OPERATIONS WITH COMMON FRACTIONS

1. 5 34/45
3. 16 3/8
5. 1 11/21
7. 5/24
9. 13 1/3
11. 3 3/4
13. 13/24 yd
15. 3/4 lb
17. 50 1/8 in
19. 32 3/8 in
21. 11 1/6 ft
23. a. 5 5/8 lb
    b. 1 7/8 lb
    c. 3 3/4 lb
    d. 11 1/4 lb
25. 4 1/4 in
27. 160 sheets
29. 618 in

SECTION 3 DECIMAL FRACTIONS

UNIT 11 ADDITION OF DECIMAL FRACTIONS

1. 574.63
3. 393.633
5. 1 249.4 cu in
7. 145.32
9. 1 758.076 3
11. 825.152 lb
13. 199.5 psig
15. 0.625 in
17. 92.3 Btu/lb
19. 0.25 lin
21. 3.18
23. 62.8 lb
25. 12.72 amp

UNIT 12 SUBTRACTION OF DECIMAL FRACTIONS

1. 123.42
3. 1.294
5. 42.767 sq in
7. 89.8
9. 618.015
11. 4.257 3 m^2
13. 6.858 lb/cu ft
15. 9.464 lb/cu ft
17. 0.143 in
19. 0.0186 lb/sq in
21. 1.212 in
23. 34.7 lb
25. 727.49 psi

UNIT 13 MULTIPLICATION OF DECIMAL FRACTIONS

1. 2 251.092
3. 5.486 112
5. 7.568 739 lb
7. 3.507 063 7
9. 0.011 084
11. 40.930 7 cm^3
13. 0.405 lb/sq in
15. 14.62 lb
17. 42.687 5 Btu
19. 876.174 Btu
21. 13.3114 amp

UNIT 14 DIVISION OF DECIMAL FRACTIONS

1. 3.52
3. 0.418
5. 7.522 gal
7. 531.5
9. 3.082
11. 33.35
13. 1.835 amp
15. .0028lb
17. 1.65 gal
19. .284 9
21. 68.86 Btu
23. 0.097 in
25. 2 116.8 cu ft

UNIT 15 DECIMAL AND COMMON FRACTION EQUIVALENTS

1. 0.187 5
3. 0.875
5. 0.175
7. 0.266 7
9. 0.166 7
11. 2.6562
13. 439.74
15. 467.749
17. 1.102 in
19. No
21. 1.379 in
23. $46.71
25. 3.32 gal

UNIT 16 COMBINED OPERATIONS WITH DECIMAL FRACTIONS

1. 20.886 1
3. 12.265
5. 11.085 09
7. 4.52
9. 90.208 8 cm
11. 7.893 ft
13. 19.162 2 m
15. 2.539
17. 0.375
19. 0.15
21. 1.61 in
23. 91.597 5 lb
25. 0.251 lb/ft
27. 2.074 8 in
29. 8.223 3 in

SECTION 4 PERCENT, PERCENTAGE, AND DISCOUNT

UNIT 17 PERCENT AND PERCENTAGE

1. 0.02
3. 0.12
5. 0.837
7. 4.15
9. 143%
11. 160
13. $456.40
15. 6 h
17. $24 000
19. 3.6 lb
21. $200
23. 33%
25. a. $1 944
    b. $174.96
    c. $2 118.96
27. $0.52

UNIT 18 DISCOUNTS

1. $46.73
3. $506.25
5. $992.64
7. a. $3 515.50
   b. $3 752.50
9. $208.18
11. a. $388.37
    b. $406.22
13. $2 932.16
15. $160.13
17. $12.69
19. $0.66

SECTION 5 RATIO AND PROPORTION

UNIT 19 RATIO

1. 1/2
3. 5/1
5. 3/2
7. 3/7
9. 4/3
11. 11/7
13. 7/9
15. 9/11
17. 7/22
19. 61/64
21. 11/4
23. 10/7
25. 7.85/1
27. 28/3

UNIT 20 PROPORTION

1. 15
3. 18
5. 27
7. 11 2/3
9. 1 597.5 lb
11. 7.2 ft
13. 12 960 cu in
15. 0.021 in of water
17. 850 rpm
19. 7 .5 tons

SECTION 6 DIRECT MEASURE

UNIT 21 UNITS OF LENGTH MEASURE

1. 60 in
3. 86 in
5. 175 cm
7. 7 cm
9. 5 ft
11. 4 1/6 ft
13. 4.65 m
15. 5.50 m
17. 31 in
19. 45 in
21. A:12 1/2'×12 3/4'
    B:10'×10 1/3'
    C:10 1/6'×13 1/2'
    D:9 1/6'×14 1/4'
23. 31 in
25. 1 046 cm

UNIT 22 EQUIVALENT UNITS OF LENGTH MEASURE

1. 22.86 cm
3. 48.26 cm
5. 2.13 m
7. 5.906 in
9. 2 ft 9.858 in
11. 22 ft 11.591 in
13. 15.24 cm
15. a. 9 ft 10.1 in
    b. 16 ft 4.9 in
17. 9.842 5 in
19. 11.647 2 m
21. a. 60.96 cm
    b. 121.92 cm
23. a. 50.8 cm
    b. 40.64 cm
25. 4.72 fins/cm

UNIT 23 LENGTH MEASURE

1. 19' 10"
3. 16 ft 10.5 in
5. a. 1 040 cm
   b. 10.40 m
7. 26 ft 6 3/4 in
9. a. 1.23 m
   b. 0.67 m
11. a. 6.55 cm
    b. 8.1 cm
13. 9 ft 9 in
15. 48 in
17. 56 in
19. 30.27 in
21. 50 3/4in
23. 84 in
25. 2.513 in

UNIT 24 ANGULAR MEASURE

1. 35°
3. 127°
5. 319°
7. 30°
9. 85°
11. 45°
13. 61°
15. 45°

UNIT 25 EQUIVALENT UNITS OF TEMPERATURE MEASURE

1. 25°C
3. 77.8°C
5. 185°F
7. 348.8°F
9. 393 K
11. 296 K
13. 10°C
15. 20°C
17. a. 18.3°C
    b. 60°C
19. a. 68°F
    b. 37.4°F
21. 4.4°C
23. –15°C
25. –6.7°C

SECTION 7 COMPUTED MEASURE

UNIT 26 AREA MEASURE

1. 64 sq in
3. a. 0.562 5 sq in
   b. 25.312 5 sq in
5. 80 sq ft
7. 960 sq in
9. 6.16 m²
11. 38.485 sq in
13. 0.64 sq in
15. 84 sq ft
17. circular
19. 12 in

UNIT 27 EQUIVALENT UNITS OF AREA MEASURE

1. 12.5 sq ft
3. a. 375 cm²
   b. 0.037 5 m²
5. 91.5 sq ft
7. 8 300 sq in
9. 53.014 5 sq in
11. 464.52 cm²
13. 248 sq in
15. 12.9 sq ft

UNIT 28 RECTANGULAR VOLUMES

1. 720 cu ft
3 $0.3\ m^3$
5. a. 6 415 1/2 cu in
   b. 3.71 cu ft
7. 2 807.5 cu ft
9. 1.75 ft

UNIT 29 CYLINDRICAL VOLUMES

1. 2 463.014 cu in
3. 49.088 cu in
5. 0.136 cu ft
7. $243\ 698.589\ cm^3$
9. 214 cu ft

SECTION 8 FORMULAS

UNIT 30 OHM'S LAW AND ELECTRICAL RELATIONSHIPS

1. 2 Ω
3. 2.5 A
5. 2.67 A
7. 9.58 Ω
9. 0.75 A
11. 2 990 W
13. 227.27 V
15. 10 Ω
17. 10.29 Ω
19. 100 μF
21. a. 11.5 V
    b. 129.2 Ω
    c. 288.4 ft

UNIT 31 GAS LAWS

1. 12 cu in
3. 23.33 psia
5. 11.60 psia
7. –96°C
9. 59 psia
11. 27°C
13. 57.6 psia
15. 0.11 cu in

UNIT 32 HEAT LOAD CALCULATIONS

1. 4 145 Btu/h
3. a. 4 529 Btu/h
   b. 79.5%
5. 5 595 Btu/h
7. a. 2 068 Btu/h
   b. 3 367 Btu/h

SECTION 9 STRETCHOUTS AND LENGTHS OF ARCS

UNIT 33 STRETCHOUTS OF SQUARE AND RECTANGULAR DUCTS

1. a. 48 in
   b. 18 in
3. a. 32 in
   b. 26 3/4 in
5. a. 90.8 cm
   b. 75 cm
7. a. 108 1/4 in
   b. 24 3/4 in
9. a. 37 1/8 in
   b. 36 in
11. a. 121 1/8 in
    b. 30 in

UNIT 34 STRETCHOUTS OF CIRCULAR DUCTS

1. a. 18 7/8 in
   b. 24 in
3. a. 75.40 cm
   b. 50 cm
5. a. 94.95 cm
   b. 100 cm
7. a. 24 11/16 in
   b. 28 in
9. a. 71.52 cm
   b. 70 cm

UNIT 35 LENGTHS OF ARCS OF CIRCLES

1. 1.57 in
3. 15.71 ft
5. 150°
7. 4.19 cm
9. 21°
11. 3.14 in
13. a. 3 15/16 in
    b. 11 3/4 in
15. a. 3 9/16 in
    b. 10 5/8 in